LEVOLUTION

LEVOLUTION

COSMIC ORDER
BY MEANS OF
THERMODYNAMIC
NATURAL SELECTION

MICHAEL GUNTER

ARCHWAY
PUBLISHING

Archway Publishing books may be ordered through booksellers or by contacting:

Archway Publishing
1663 Liberty Drive
Bloomington, IN 47403
www.archwaypublishing.com
1-(888)-242-5904

ISBN: 978-1-4808-1008-2 (sc)
ISBN: 978-1-4808-1007-5 (hc)
ISBN: 978-1-4808-1009-9 (e)

Library of Congress Control Number: 2014914882

Printed in the United States of America.

Archway Publishing rev. date: 09/30/2014

CONTENTS

PREFACE

The personal discoveries that led to this book started to come to me about 30 years ago, while I was trying to write in my spare time about the parallels between ecology and economics. After a degree in Biology at Kansas University, and a Masters at the University of Texas, my academic grounding was centered on evolutionary and marine ecology.

At the time I started the book I was immersed in environmental consulting related to Houston's oil and gas economy, writing parts of environmental assessment reports for large industrial projects. It was during this time when I sought to get a better understanding of the "soft" sciences of economics and sociology, and to integrate into these, the two "hard" concepts of thermodynamics and evolution. It wasn't that hard at all. Our culture, our economy, and our ecosystems all run on energy. So, in 1983 I felt like this topic was at the nexus of everything important.

As I continued to ponder it all, the scope of phenomena and systems that I considered as compliant with my emerging theory about a new kind of evolution involving new levels of organization, continued to expand. After more time, it covered so much ground that I began to see that I was on to something very grand in scope. It all went to my head, and I have not been the same since.

Then I finally heard the news from physical chemistry about how Ilya Prigogine had uncovered "dissipative structures." This lent considerable support to the alternative concept of "order" that I was developing. I then began to realize that the ecology- and evolution-based story I was weaving could actually find a home in almost any discipline. Why would that be true? I pursued this question, and the answer turned out to be elegant and simple, and led to this book; these phenomena are universal.

This book then, is about the universe in its generality, and at the most fundamental levels. Observations have resulted in our perception of regularities or laws, and the most fundamental of these are the laws of thermodynamics. Levolution represents and proposes an expansion of the classical laws of thermodynamics. It proposes seven new laws for a total of ten, so they will now fit perfectly on two clay tablets. Appendices B and C constitute a list of them.

I am very happy that I finally got this book done, but I want to convey substantial humility about the lack of mathematical support. I was lucky enough to receive the revelations, but they did not come with any formulae.

I hope this book is sufficient to pass along these assertions that I think are very important to the advance of science, and give the reasonably well-educated layperson a view of the grand and universal process of Levolution, which has given a simple kind of understandable order to the universe.

Toward that end, I have kept the description of this simple theory quite simple. I have not found it necessary to involve mathematics to get the points across. The objective here is to convince readers of the reality, scope, causes, and characteristics of Levolution. One should at least get an understanding of how this very general theory works, with enough evidence to convince you that it is quite real. Researchers in the appropriate disciplines will need to fill in many more details, but I am confident that they will.

I have taken pains to maintain a simple and understandable approach to two confused areas, entropy on the one hand and evolution on the other. These important subjects are at the very core of Levolution. I have had to coin several new terms in order to be precise where it counts, but if one accepts these few neologisms, the payoff will come in the form of clarity. There will be no quibbling about where the theories may be applied.

This book is about what I seriously consider the most important natural process in existence. There would not be gravity, stars, or life without it. Levolution has created nearly every kind of naturally occurring system in the universe. It is based on the simple properties of energy flowing in systems, yet it has built the contents of the universe.

The book is planned as the first, foundational installment of a series of related works, but because they are not all finished, I will simply tease you with the notion that more is on the way, and a whole descriptive *Levolutionary Cosmology* based on this book's foundation of principles is due at the publisher soon. The whole exercise has been a great success, and the cosmology based on the principles of this book seems to open vistas for future science to explore.

The revolutionary aspect of these ideas will predictably attract some believers and some non-believers. There will be those who will attack, and there will be those who accept them. I therefore include a plea for support from the believer's side. These early concepts beg for a research program designed to nail them down further. I can envision an institution where targeted research along these lines can be funded, and visionary philanthropists can make a difference in the world by actually promoting this new and unified science of how the universe came to be.

Thanks, Credits, and Acknowledgments

I really did find my initial inspirations for this work long ago in the words of the ancient Greek philosopher, Heraclitus. I honor this fact by replicating his words at the head of every chapter here. It is almost haunting how well he presages my chapters with one-liners. They are just a few of the translations of Burnet (1930) and are listed in Appendix A.

I would like to thank Professor Ed Wiley, my former Kansas University zoology professor, whose presentation vigorously defending Evolution against the attacks of the Intelligent Design crowd was most impressive. That was in the context of a "debate" during my time there at KU. Wiley, and co-author Daniel Brooks, have contributed to my thinking in this endeavor directly. Their 1988 book, *Evolution as Entropy*, was on the path.

I was formally schooled in the works of the giants of biology; the visionaries like Darwin, Bertalanffy, Lotka, Hutchinson, and the Odum brothers, early in my biological career. Many discussions about fish energetics with my major professor, Donald E. Wohlschlag, helped to integrate their contributions under the banner of energy, and I owe him my gratitude for giving that proper emphasis.

The notions of many great scientists are here. This work has been a great education for me. I thank all my many teachers. I feel like Levolution would certainly be different if I had not encountered on my own, the integrative works of Ervin Laszlo, Pierre Tielhard de Chardin, Ken Wilber, Kenneth Boulding, Talcott Parsons, Richard Dawkins, Stuart Kauffman, and Lee Smolin. This list does not name many other influencers, but they will understand.

Thanks also go to my wonderful wife, Patricia, a hard-working physician who "has not ever had any shred of interest in this stuff" for providing the passive resistance that made me want to make this work more accessible and interesting. Her good work also afforded me the comfortable time in Hawaii to get it finally done. I also thank Patricia McCoy Greenwell, a former colleague at UT, who actually coined the term "omniculture," which I used. She, along with friends Lance Caspary, Iain Tyson, Zack Shamm, and Kathleen Dunne, all contributed to proofreading the manuscript and made good comments. I must also give a big Thank You to the Wikipedia Project, which makes for very easy access to the ideas of history's geniuses.

CHAPTER 1

ENERGY IN ALL THINGS

It is wise to hearken, not to me,
but to my Logos, and to confess
that All Things are One.
— Heraclitus

INTRODUCTION BY HERACLITUS

So begin the fragments of the ancient Ionian Greek philosopher, Heraclitus, whose reasoning, which he called his Logos, was very sound as he described his early cosmology about 2500 years ago. All things really are one, as I will tell it as well, and it has been since the beginning of time. Heraclitus is most famous for describing how "Fire," and that is "ever-living Fire," is the primal substance. The stuff from which all things are made is actually energy, but Heraclitus' conception is acceptable as a general reference to energy.

With the "ever-living" adjective, Heraclitus even nails the first law of thermodynamics, the conservation law. Energy really is "ever-living" because it is never destroyed. Energy is always flowing over time, causing what we call change in the process. It is the cause of all change in fact, and is involved in everything that happens. It is at the base of all things. You do not have to take that on faith. It is human-derived science. There has never been a contradictory observation. It is accepted by virtually everyone, as one of the more profound truths about our universe. It may not be such an amazing thing that Heraclitus came up with this particular scientific truth so long ago. The ancient Greeks were perhaps more advanced than we generally appreciate. Heraclitus was sufficiently confident to take an unmistakable tone of certainty in his aphorisms.

It has not escaped me that if one replaced the word "energy" with the word "God" then that deity would be in the same logical position as the subject of thermodynamics. Thermodynamics are the laws of energy and change. I mean no disrespect to anyone, but that is not the direction we will take here. This is a work of science.

Heraclitus is known for his views on change. He gives us a far ranging description of energy's operation in the universe, discussing its potential between opposites, or tension, its lawfulness, or justice, and the transformation of its very nature. His words are dated, but they are not exactly wrong if you give him some latitude. As if that were not enough, Heraclitus's logos, his reasoning, is also sound about the "oneness" of all things. Physics today searches for the details of "unification" among the many forms of energy that have appeared since beginning of time.

Although I have never heard any of his interpreters say it, having lived with the profound fragments of Heraclitus for over 30 years now, I can make a good argument that Heraclitus also understood the mechanism of change we know as *natural selection*. It is what he called "the kingly power," and relates directly to justice. He understood the process as a simple and ubiquitous situation; the simple reality of life and death.

This reasoning is his mechanism for spontaneous change. Change is a given in the philosophy of Heraclitus. In a simple view of it, and in his own fragments, change is essentially by "thunderbolt." The Ionian Greeks in 500 BCE were at the beginning of science. They mixed science and religion. Heraclitus himself said that Zeus both does, and does not, care what you call him. The image of a bolt-wielding deity is sufficient to deliver my meaning. One of his statements is "The thunderbolt rules the course of things". As we shall see, it is really *Thermodynamic Natural Selection* that rules the course of things throughout the universe.

There are many things to like about this philosopher's delivery in a powerful style of wise-sounding, one-line aphorisms, and especially the notion that these ideas are still around and relevant today. They actually seem to be the roots of this tale of Levolution. They actually were the roots of my own intellectual adventure here. I will be channeling Heraclitus throughout the book.

Energy's flow in the universe is what this book is about, almost exclusively. Energy's flowing animates everything in the universe. Nothing happens at all without energy. All things are truly one, mainly because all things are ultimately made of energy. Energy's story is the story of the universe, and each one of us is an energetic entity, and a part of the story while we live. Most importantly, the universe is all one big story. Science currently accepts that it has a big bang beginning, a universal entropy project as a plot, and entropic doom as an ending. This book is mainly about the plot. This plot is governed by the laws of nature, began at the very beginning, and has not changed, even as the actors have developed and differentiated.

Heraclitus seems to originate the threads of a cosmic reasoning, his logos, that I can tell has continued through the history of human civilization, and is found today in both science and in multiple major religions. I quote Heraclitus at the opening of each Chapter here, and I have compiled the collection of these quotes as Appendix A. I simply pass along the translations of Burnet (1930).

Levolution: A New Paradigm About Nature

Levolution is defined and explained herein as a newly discovered, energy driven process. It is the process of changing a group or population of existing entities such that they organize into a single new whole, a new entity, a new monad, or new kind of unit. Levolution is the science behind the "organization process" of nature itself. It has built the levels of organization of the universe; all of them. The things in the universe do not self-organize any more than they self-evolve.

The organizing is always done by the same process, and its causes are always external to the things being organized. The simple fact is that Levolution is caused by natural selection. This universal and cosmic process, mentioned by Heraclitus and discovered for science by Darwin, is behind both evolution, which is adaptive change, and Levolution, which might be called integrative change. I capitalize the word Levolution throughout the book.

The integrative change of Levolution is so central to the development of the universe that it is no exaggeration to say that Levolution is the process that has created virtually every material thing we know. Think about it a little and you will realize that virtually every natural thing is composed of parts. For a given subject entity, at some time in the past, (a) its true parts were once independent wholes in a population, and (b) they are now organized into the whole that is the subject.

Not quite everything exhibits this pattern, but the list of things that do, shown listed in Figure 1, plus the names we give to variants of them, and to pluralities, populations, or groups of them, do add up to just about everything. To the things that do exhibit the pattern, I will have to give a new name. They are *holosystems*.

I have discovered the *holosystem* or the *holosystem structure*. These structures are all similar in that they all exhibit the same schematic or generalized pattern of energy flow. They all do essentially the same thing from the perspective of energy. I will relate observational evidence that the process of Levolution has created the first example of each holosystem. Their parts were once free and independent entities, and now they are not.

I should note here that holosystems are actually a special subclass of a known thermodynamic form, called the *dissipative structure* that was discovered by Ilya Prigogine in the 1970s. He won the 1977 Nobel Prize in Chemistry for the discovery, and this structure or pattern is now recognized as the foundation of non-equilibrium thermodynamics, and its growing sub-discipline, morphodynamics.

Dissipative structures are always discrete entities or systems that manipulate a flow of energy into an internal order or a pattern. Swirls at drains and hurricanes are both examples. Like a wave in water, they are not the water itself; they are merely the pattern of energy in the water. A water wave, in fact, is a morphodynamic structure of a slightly different type.

Figure 1—List of Subject Holosystems

ELECTROMAGNETIC AXIS

0. Photons and Gluons
1. Gravitons
2. Quarks
3. Nucleons
4. Nuclei
5. Stable Atoms
6. Molecules and Chemical Reactions
7. Autocatalytic Reaction Sets
8. Prokaryotic Cells
9. Eucaryotic Cells
10. Multi-celled Organisms
11. Societies / Structured Populations / Colonies
12. Ecological Communities / Multi-species Cultures / Civilizations
13. Ecoregions and Biomes
14. Planetary Ecosystems

GRAVITATIONAL AXIS
(in Mass Order)

15. Dust (Atoms & Molecules from Levels 5 & 6)
16. Planetesimals (Comets, Asteroids, etc.)
17. Moons
18. Planets
19. Stars
20. Galaxies

DARK ENERGY AXIS

21. Dark Gravitons
22. Early Dark Matter in the Voids
23. Later Dark Matter in Filaments and Clusters

The special things about the dissipative structure pattern are that it forms a discrete structure, or a system in an environment, and it increases energy throughput, all while creating its own order. Increasing the energy flowing through a discrete dissipative system always also increases what physicists call the overall entropy production of the system.

This point is a little bit technical, but highly important, so hang on to it. The creation of its order requires internal work, but even the useful, order-creating work done by such a system eventually becomes an increase in the entropy produced by the system. Even order-creating work promotes the degradation and dissipation of energy in the universe. This is simply how energy works, and we have known this since Sadi Carnot's "heat engine" was explained in the early nineteenth century.

A dissipative structure, an energy-flow schematic of which is diagramed in Figure 2, may be composed of anything that allows energy to flow within its medium. In swirls at drains and hurricanes, the medium of energy flowing is usually a fluid, like air or water, experiencing gravity and the raw forces of nature.

Figure 2—The Dissipative Structure

Energy Use,
Internal Ordering
and Work

**Energy
Capture**

**Energy Use,
External
Work**

**Energy Degraded and/or
Dissipated as Entropy**

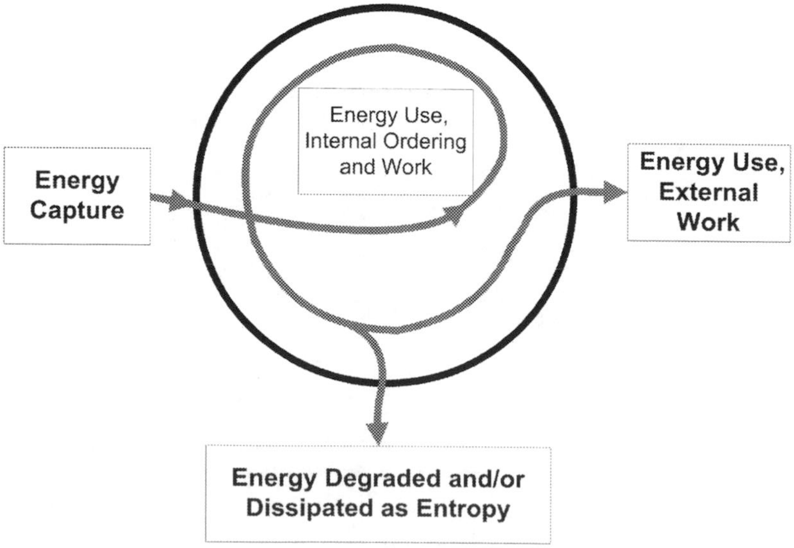

In holosystems, which are a true subset of the dissipative structures, the medium is not a fluid, but a population of previously existing parts. Each entity serving as a part of a holosystem is also moving energy in a pattern; the very same pattern as the whole. This leads to my definition of the holosystem.

The holosystem structure is a special class of dissipative structure that (1) is always composed of parts that are also dissipative structures, and (2) has formed by Levolution.

Looking at the universe with a particular slant, the slant created by an understanding of holosystems and the process of Levolution that makes them, yields a new view about how nature fundamentally works.

It is not a mysterious interpretation, but it is profound, and its concepts lay at the roots of many belief systems. I want to give you some recently discovered scientific facts, and some new arguments and perspectives from the distant past. These will complete a gestalt or holistic view of a very different universe than you may have contemplated.

Levolution is a scientific subject, but it is a very generalized or abstract explanation of what the universe is doing, what it came from, and how it has created all the entities that we observe. That it leads to a unification of our understanding of nature is surprising, but the fact that it leads also to conceptual unification of the various forms of energy is a surprising bonus. Levolution will unlock a robust Theory of Everything.

Levolution, as it's facets unfolded through my reading, writing, and curiosity over 30 years or so, has grown into something of a Grand Theory. It has found its scientific grounding and become a sweeping paradigm that first boldly revises the laws of thermodynamics, and then proposes a related new set of natural laws to explain the growth of what I will call *functional order* in the universe.

Best of all, the effort has proven that science, and the picture painted by its discoveries, can yield a sense of meaning for the universe as a whole, and our place within it. It is not surprising to me that the meaning is not centered on humanity, as so many historical religions, and certain quarters of physics, believe. Nor is it centered on achieving permanence or immortality. Levolution gives only the solace of nature's style of wisdom, but the meaning given is clear, it is defensible and real, and it should motivate us to work hard, and to facilitate the growth of functional order.

Work is the expenditure of energy by a system doing something that is useful *to it*, and in support of the *entropy project* that the universe is performing as a kind of plot underlying the drama. We can now work hard and work smart, with knowledge of our mission, because it is the same as that of anything else.

Work for yourself, your family, your society, your ecosystem. Have some expectation that your work will provide what is needed for you, your cells, their reactions, their molecules, and their continued chemical arrangement. This simple statement covers the major part of the profound meaning I have gathered from this new cosmology. It has a familiar ring to it. It is perhaps the case, however, that different people will find differing profundities. I feel strongly that the raw materials for many systems of knowledge and belief are within these pages of scientific explanation.

To understand this more fully, consider that the universe is all about flowing energy and that energy has a property, somewhat like gravity, that makes it always go in one direction, downward in potential.

Work takes some quantity of energy downward in potential, as do any non-useful energy flows. Work done internally, however, means that some of it was "siphoned off" and used to create internal order in some dissipative system.

That order is what we must call *entropically functional*. This newly discovered kind of order, it turns out, is created for the very purpose of making energy flow downward even faster. Functional order does this by virtue of the special pattern of energy flow in a dissipative system. You are one of those systems.

Consider also that energy is a lot like water flowing down a hillside under the influence of gravity. The topography reflects the hill's many detailed constraints, or its particular order, and its valleys and channels will vary in steepness, width, and depth. Most of the water coming down, like most of the gravitational potential energy it represents, will take the steepest available channels downward. *It will automatically allocate itself to the pathways that move it downward the fastest.* This simple principle, we will see, is a long-overlooked principle of thermodynamics, discovered over thirty years ago by Rod Swenson.

This book explains the Laws of Energy and Functional Order, and reveals the two special structures we have already mentioned. It also covers three very familiar causes of change among energetic systems.

A Greek Diversion

The Greeks had some of this figured out, I am essentially re-writing and updating Heraclitus's Logos, and we are in a position to now understand it, perhaps for the first time in a long time.

Heraclitus' Logos, or reasoning, "had legs" and it was picked up by Plato about a hundred years after Heraclitus. Plato's ideas became a part of Aristotle's conceptions of the Four Causes. They also became Platonism, and found homes in Neo-Platonism and Stoicism. The executed prophet, Jesus, with a little help from Constantine passed them into Christianity, and Western Philosophy.

In the important philosophy of Plato, there is an important distinction to be made between our world, and an alternative, higher realm of purity and perfection. To Plato's mind there is a realm of perfection above us, (where the revolving stars make perfect circles) a realm of Ideals, and Ideal Forms, and the diverse, ethereal, and separate "Forms" of All Things. Aristotle would connect with this and transform it into his Formal Cause, one of his Four Causes, which we might recognize today as simply a template, design, or plan.

In Plato's ideal realm, there was a Master Ideal Form, the "Eidos" and all the other Forms of various things, the "eide" that exist only in this Ideal realm. This whole region is separate and apart from we humans and our world, but we do have a role. We mortals do not enter into the Ideal realm alive, but the Ideals and Forms of that world are shown to us because they are intelligible. We can sense them. We know the form of a good table, for example, when we see one.

In Plato's view, we, and everything in our real world, try to emulate the Ideal Forms. We fall short, of course, and are imperfect, because we live in the material, Heraclitean world of flux and change. Here we are judged, and subjected to "the kingly power." We atone for our sins, and we hope to avoid the thunderbolt. In short, we are subject to death; we are mortals.

Plato suggested that the things in our world "participate" in the Ideal Forms, to make our worldly reality, but they do so imperfectly. The participants are the physical objects or matter in the universe which instantiate, populate, or mimic the Ideal Forms, but do so down here in our reality.

The things that are real do not get into Plato's heaven-defining, ideal realm without giving up material, which is part of the source of the imperfections. The "Ideal" to Plato is akin to perfection, like "The Good", but the material things here in our world are representations of ideals. Shape is given to participating matter by virtue of its *eide* or its particular Form. All of the eide, or diverse forms of things, stem from the divine source, the Eidos, or the Master Form.

These are radical departures from our current ways of understanding. They are far from current science, but they are not unfounded in either reality, or in science. Plato may have been closer to correct than most people know. You see, I think the dissipative structure of thermodynamics could be interpreted as the Ideal Master Form, the Eidos. The holosystem is, likewise, a major form or structure, an "eide" derived from it. The things subject to Heraclitean change today are the things we can categorize as dissipative structures and holosystems.

In the long run of universal and cosmological development, three phenomena seem to cause the real change among the things in the universe. Here I mean the major changes that are really changes in Form, where one thing actually becomes another. Such changes, perhaps all such changes, are related to energy and order, but they have not come to me from either physics or philosophy. The subjects here are the laws that cause change. These are just observed regularities, but they are also another kind of ideal. They are (1) Thermodynamic Natural Selection,

(2) Universal Thermodynamic Evolution, and (3) Levolution, and this book is about them. The Ideal Forms and the means to change them are herein presented as the Laws of Functional Order.

To Plato, these particular natural laws would be a connection between worlds. They are the means to change forms down here in our reality. They would seem like communications of "design" from the realm of the Master Ideal. Other thinkers have contemplated these things, and they all use slightly different analogies. We will take a principled view based on current science.

Strangeness of the Content

This treatise is aimed at anyone curious enough to deal with the abstract nature of the many interacting hypotheses and theories here. It is written casually, rather than formally, but I do not protect the reader from the details that support the logic. No complex mathematics obfuscates the exposition here. References are provided for those who want to dig in. I am walking a thin line here and I know it. I want these ideas to reach the halls of academia, but I do not normally walk those halls. A work of popular science seems like a reasonable way for an outsider, an external change agent, to get word of Levolution out to science and the public.

A divergent framework for important scientific facts, theories, and ideas is a paradigm. It is a paradigm whether or not the facts and ideas are true or false. If it serves as a framework for a set of connected ideas, it is a paradigm, whether or not the facts are all in, and whether or not some of the ideas are speculative. Without qualification then, Levolution is a clearly a paradigm, a radical shift from normal science. The Levolution Paradigm is an extension of thermodynamics into the realm of functional order. It's veracity is something else.

This foundational book of a planned series contains newly proposed natural laws. In it, my assertions are not generally speculative, but the perspectives will lead us to a new look at the universe. When we use the paradigm of Levolution to look again at cosmology, we will need to speculate. Science has yet to figure it all out, but it is getting close to a new gestalt, and Levolution will contribute to that.

For many people, the various associations here will seem a bit strange. They are explanations of the observed regularities associated with how the universe operates. We are not used to thinking so deeply about parts and wholes, for example.

Levolution, this book, is only the beginning of a work of science. It begs some questions, but it organizes a new and improved picture that links many profound, proven, and accepted truths. This book is the foundation. Levolution is based on sound thermodynamics, and a small revolution related to natural selection that has been trying to happen since Alfred Lotka wrote a paper about it in 1922.

I think Levolution shows a way forward, but there is much work to be done. The ideas do create a framework, and the framework is novel and revolutionary. The next steps will aim to delineate the disprovable hypotheses, and test them, but here in this book, the aim is to make the logical connections to thermodynamic law that will build out the framework.

While I try to reach anyone with a general education, the subject here is broad and approachable from many directions. It is difficult, but rewarding. I am totally convinced and serious about the science of Levolution, but this work is one of "popular science" and the audience targeted is broader than any one discipline can train. It is designed to hold the attention of physicists and cosmologists, but also lay people, and the cosmically curious.

I take some potshots at physics. I am not impressed by physicists who believe the unbelievable without sufficient cause. I am not happy with where physics is going. The interpretations of quantum theory have led physics to an unbelievable reality, and seeks to cause science to abandon causality, and apparently seeks to change the game of science itself. All this is done in its reverence for overly complex mathematics and a misinterpretation of one morphodynamic experiment involving slits and waves.

Physicists like to say they are going where experiment leads them, but in reality, and for eighty years, they have gone only where one mistaken Danish interpretation by Max Born has led them. Physicists are today even seeking support for the idea that the universe is here for humans, the so-called Anthropic Principle. I will rest my case for now.

It is clear to me that there are some academic battles ahead, but we must collectively fix and expand the Laws of Energy and Functional Order if we hope to progress in our understanding. I implore the reader to suspend disbelief for a while. It all comes together eventually.

In this foundational work on Levolution, we will ascend step-wise though the growth of functional order, built up by the process of Levolution, even as the one thing in the universe, the energy, degrades to weaker forms, similarly step-by-step through entropy's continual increase.

We will contemplate the universe's long history of spontaneous production of novel systems on levels of increasing order. The mythology of what is widely being called *self-organization* will be addressed, and easily refuted.

In the Levolutionary development of material structures in the universe, we will look briefly at all twenty-three levels of material organization, and note how each system, on each new level, degrades energy further, and moves more energy downhill because it is larger in scale. We will avoid any spiritual interpretations, simply because this is a project of science.

Levolution is a new paradigm that provides an intellectual twist to nearly the entire framework of current science. It is based on several non-traditional, but easily-proven principles of which I am confident, and it yields non-traditional explanations. It is not a set of proven cosmological theories, but is a framework of statements of "first principles" that also happen to all work together, producing a kind of logic.

The actual observations that are most crucial to the case for Levolution have already been made. What are missing from the perspectives of current cosmological physics are mainly found in the principles of the biological science of ecology. Given these, plus a clear-headed logic, and an energy-centric perspective, Levolution takes a central role in cosmology, just as Darwin's Evolution has taken a central role in biology. Levolution stands as an offered explanation, based firmly on natural laws, for the universe we already know, but it also provides some explanations for the universe that we don't know.

Unlike Heraclitus, I will apologize in advance for any seeming overt boldness. To put Levolution where it belongs, I have needed to take some major liberties with accepted physics, and reiterate some important new natural laws, most discovered by others.

The main observation to be explained by Levolution, however, is the profound observation that the universe is "nested". I refer to this "nestedness" not as "hierarchic" but more accurately as the "holarchic" structure of the universe; to emphasize the sequential levels of organization of parts and wholes, rather than a system of command and control. The word and the concept I have borrowed from Arthur Koestler's *The Ghost in the Machine* (1967).

Levels in the Universe

Aristotle's conceptions of the universe include his famous description of its "beings," represented as a series of levels. It is a linear taxonomy extending downward from perfection (the Platonic Ideal Realm) of the divine. The Great Chain of Being that developed from his conceptions was the beginning of the Western idea that not all of the things in the universe are even of the same realm.

Between Aristotle and the much later Scholastics of the Middle Ages, the universe as it was seen in the West, gained a deeply layered structure. Heaven, Deity, and Angels were at the top, while Satan, Demons, and Hell were at the bottom. In the middle were material, earthly things like us, and we are up closer to Angels. Below us are the other "blooded" animals, and below them are simpler "bloodless" animals, which systematic biologists now call the phylum of Invertebrates. It was a start.

Aristotle's ideas about the levels and the Great Chain of Being create a picture of an ordered universe with a layered structure, in which things have their place. It was perhaps the first one. That picture is near the beginning of the story of cosmic order that I will relate in this book.

Levolution as a Process

The process of Levolution created the levels of organization in the universe. Levolution is a thermodynamically driven process that results in new energy-flowing systems on a new, and higher, level of order. As I have mentioned already, there are two related phenomena that cause and support Levolution.

The first is the famous and profound mechanism of natural selection, which is actually thermodynamic in its nature and its origin. It emerges as "differential demise" among variant types in a set of discrete entities. The second supporting actor, allowed by the mechanism of the first one, is the universal version of Darwin's adaptive evolution. These are the apparent primary tactics of energy in producing adaptive change as needed among its discrete energy flowing systems. Together these two linked processes select, form, and continue to adapt all the natural systems to their changing environments. They are adaptations of energy to time and change.

The difference that Levolution adds to Darwin's theory is, at its base, very simple. It is the notion that the *environment* of the evolving energetic systems of the universe, the source of natural selection itself, is very likely to be one of two things. It is either (1) the inside of the system on the level of organization just above it, or it is (2) an unordered region containing other systems, where energy flows may yet become ordered by spontaneous formation of a higher order structure, through an event of Levolution.

In other words, environments either are aspects of the next level up, or are regions where such might form. If you are not a scientist, there is nothing really incorrect or wrong with failure to recognize this fact, and simply continuing to refer to "the environment" as the outside of a system, but this incomplete view fails to show biological species, for example, as the functional "parts" of ecosystems that they truly are. This general relationship may be extended universally.

Levolution is also a story of the search for the very meaning of "order" in the universe. In the ancient past, it was conceived as a Divine Order to be sure. Science will generally begin removing God from the order of the universe in the Middle Ages. By now, early in the twenty-first century, with the help of a new thermodynamics, the story can be told about energy's tendency toward *entropically functional order,* and the story is purely scientific in its orientation.

The scientific orientation is mainly to maximize believability. There is plenty of room to wonder at, and even be in awe of, energy and the picture painted by the Levolutionary Paradigm. The last book in the Levolution series will be written from a more cultural perspective, where spirituality, religion, belief systems, faith and other non-scientific concepts can be treated. As I have noted, the fodder is here for the mill of religion, and my tale here has been told before. The Logos has taken many forms over the years.

So many "scientific" mysteries remain that mysticism clearly still pulls on the minds of today's practicing scientists. People who believe in the purely mathematical nature of the universe, or those who want to believe in the reality of "quantum fluctuations" the existence of "isolated systems" or "reversible actions" are all examples of various kinds of mysticism right in the middle of accepted physics.

I am of the humble opinion that none of these things really exist. Occasional suspension of disbelief is necessary for new hypotheses to penetrate, and sometimes replace the old explanations. Try to be open-minded until you get the big picture here. Mysticism shrinks away from places where some part of underlying reality becomes understood. I am not a mystic at all, but the subject matter of "order" even as a scientific concept, is surrounded by mysticism.

We must therefore be very careful. I will dispel such mysticism as needed, but Levolution is no attack on either science or religion. Different people will get different things out of the tale I will tell. My hope is that the right people will read it and carry it forward.

The process of Levolution has left evidence all over the place, and it will not be difficult to prove to the receptive scientist, or the educated layperson, that it has occurred, and is still occurring. Levolution is however, a big story, a huge story, with that fearful word "thermodynamics" as an underpinning. It has a supportable, consistent, understandable, and believable cosmology as an objective.

Figure 1 is a list of the holosystems in the universe. We individual humans are on level ten, and our cultures are on level twelve. As noted above, *holosystem* is the name I have given to those special energetic systems (technically known as dissipative structures) that are clearly also nested and composed of parts that are self-similar dissipative systems. This schematic and structural definition has other merits, but an important one is that it guarantees that the word "holosystem" refers exclusively to systems whose origins are in Levolution.

The matter and material structures of the universe can be viewed as mere patterns of energy regardless of what level they exist upon. Einstein's famous equation, $E=mc^2$, equates matter and energy with a single constant, like a simple conversion factor. In matter, energy has been organized into fields, flows, and generally speaking, into ordered energy pathways that animate, and make discrete, the systems that it flows through. We often think of matter as static, rather than energetic, but if we were the size of an atom and could observe the various fields, we would see that matter is always full of energy flowing.

Levolution, the process that creates new levels of organization in the universe, is all about a single, special pattern related to the universals of energy flowing. It centers on that single pattern of energy flow that is involved in the creation of each of the naturally occurring holosystems in the universe.

We will examine this very special pattern in the organization of energy flows that I think characterizes both the most fundamental particles and gravitational structures as large as galaxies. The same pattern, believe it or not, is also found in the structures studied by chemistry, biology, sociology, economics, and ecology. If my observation-based conclusions are correct, a single, generalized, thermodynamically defined, and thermodynamically relevant pattern is found among entropically functional, discrete systems that represent flows of energy. The pattern of dissipative structures and holosystems represent the only way known to science that functional order can increase.

Energy may be said to be flowing in one direction only. It is on a one-way mission to lower levels of energy potential, or less energy-dense situations, and to increasingly dispersed, dissipated, or degraded forms. This simple picture takes on new dimensions when we realize that the naturally occurring systems of the universe are really only in existence because they aid in this dissipation and degradation process of entropy production.

If we look at the actual, naturally occurring structures that energy flows through in the universe, we find these are nested inside of each other like the toy Russian dolls. On one of those levels is the multi-cellular or organismal level of biology, and this is where you and I find ourselves in this schema. We fit right in, and we do the same thing that everything else does. The Levolution Paradigm provides context for explaining all human behavior, cultural development, politics, and economics as well.

THE UNIVERSE WE DON'T KNOW

Humanity exists on a tiny speck of a blue planet in a star system, in a galaxy, in a vast universe that boggles the mind in its extent. A dime held over a starless patch of sky at arm's length on a clear night covers about a hundred billion stars that you cannot see with the naked eye. The distances and numbers are vast, but observational astronomers and cosmological physicists have made tremendous progress in figuring out much of the science involved in characterizing the structures out there.

Modern optical telescopes, like the current largest ones on earth atop the looming Hawaiian mountain of Mauna Kea, have allowed astronomers to help us understand some key facts. For example, they have seen that the universe is expanding at an accelerating rate, that it contains generally the same elements that we know right here on earth, and that it is quite similar in all directions. We can also see that it has scales of structure that seem to be "nested" on a mass scale into planets, star systems, and galaxies. This is not to say that the universe doesn't still have its mysteries. There are five of them that I will mention right now.

Mystery 1 Dark Energy

The expansion of the universe is speeding up, when it seems like gravity would cause it to slow down. To account for this, physicists since Einstein have proposed possible solutions. Einstein's theory of general relativity came before the discovery of the expansion, but his first equations for gravity made the universe collapse due to gravity. He was really just trying to make the universe stable and static, as he thought it was, but his solution was to add a fudge-factor term to his equations so that the universe did not collapse in a big crunch.

He called this term the "cosmological constant" but later called it his "biggest blunder" because it was found that the universe was actually expanding and not static. Nowadays, we know that the expansion is even accelerating, and with nothing really known to cause this strange effect, cosmologists have devised the notion of Dark Energy to push the cosmos outward.

Mystery 2 Dark Matter

The outer regions of galaxies, their haloes, spin faster than astronomers think they should, based on the gravitational contributions of the visible matter there. To get the orbital velocities observed by astronomers, there would need to be hundreds of times as much matter as we can detect in the outer haloes of galaxies. To account for this discrepancy, it is now believed by most cosmologists that very large amounts of undetectable "Dark Matter" exhibits a huge gravitational effect, and must therefore exist in those regions. It even appears that Dark Matter must exist in the regions between the galaxies in the visible clusters, filaments, walls, and superclusters of the large-scale structure of the universe.

Mystery 3 The Nature of Gravity

Gravity itself represents a third major mystery of the universe. While Newton figured out how to calculate the force of gravity from the masses and distances involved, and Einstein figured out that the apparent force of gravity is really caused by distortions in the actual geometry of space surrounding massive bodies, science currently does not know how or why space becomes geometrically deformed.

This means we really do not know what causes gravity, i.e. causes the geometrical distortions and curvatures we observe. Particles consistent with the Higgs boson, have been observed. But even a Higgs particle will only a match with quantum theory predictions about its size and spin, and does not offer much of a causal explanation for gravity. A particle that directly produces gravity, the graviton, has not been observed, so even though there are ideas, there is no current explanation for gravity.

Mystery 4 The Unification of the Forms of Energy

Physicists currently lack an acceptable Theory of Everything. A Theory of Everything is technically, in physics, a theory that would explain all the various forms of energy as manifestations of a unified form. There is no current accepted understanding of why there are the various forms of energy or how they originated.

Start with the four so-called "fundamental" forms of energy:

- strong force
- weak force
- gravity
- electromagnetism

Add to these the mysteries of the singularity itself, space's so-called "vacuum energy", and then add dark energy (which is Mystery 1 here), and you have a picture of a very befuddled physics that cannot explain about 95% of the energy of the universe and draws a blank to explain the various forms of it.

Mystery 5 The Nature of Material Structure or Cosmic Order

A fifth mystery of the universe is the observable, but only partially comprehensible, structure of it. We know that at the large scales, what we do understand about gravity provides a generally adequate explanation of why astronomical objects aggregate together.

Matter, at least normal matter, attracts itself, and visible objects are consolidated into nearly perfect spheres if they are more than about 500 miles across. Some gravitational systems are in orbit around other massive spheres. We observe this general and spherical "gravitational order" to extend in scale from **single atoms** and **dust grains** to **stars** and **galaxies** at the very least.

Gravitational entities display the previously mentioned "nested" pattern among the spherical and orbiting astronomical bodies, but they do not seem to have formed via a "smaller to larger" sequence in time. Galaxies formed generally before stars, and stars formed before their planets.

The Prevalence of a Pattern

On the earth, and at our particular scale, we observe living things and note that it is "chemical order" that seems to hold the important molecules together, and arranges the reactions necessary for life. Chemical order is actually a manifestation of electromagnetic energy. Electrons in the outer shells of atoms do the chemical work of bonding to make molecules. Below in this section I will bold the holosystem names so you will get the idea of the pattern more clearly.

The ordering due to chemistry has been described by Stuart Kauffman (1993) as an aggregation of **molecules** and **reaction**s that promote each other, often in cycles. Such "**autocatalytic reaction sets**" are complex families of chemical reactions that are ordered, both spatially and temporally, in such a way that they catalyze each other. The process that brought the reactions together provides the basis for what probably led to earth's primitive **cells**.

The most primitive, **prokaryotic cells** (mainly bacteria) then are believed to have invaded either each other, and/or were consumed whole by each other. Either way, some survived and reproduced on the inside and came to live in a mode of biological mutualism called *endosymbiosis* as hypothesized first by Lyn Margulis (1970). This aggregation created what we know as modern **eucaryotic cells** with its characteristic features of membrane-bound, cell nuclei and organelles that actually share traits with bacteria and viruses.

These modern cells in turn came to live together as colonies. Colonies are often more than just pluralities, groups, or populations; they are **structured populations** or **societies**.

The cell types in these colonies then differentiated under the integrating and complementary influence of a special pattern of energy flow through the colony as a whole. They differentiated into what would become the tissue types and organs of **multicellular organisms**.

Such organisms, like familiar ants, bees, dolphins, and humans also came to live as **societies**, and all living things became organized into **ecological communities**. On earth, the sun provides the energy for most ecological communities, and so they nearly all have functional components related to the capture of solar energy; the primary producers. These are the basis for survival of the herbivores, and carnivores in turn.

The primitive types of cells are still around in large numbers, some living as independent, free-living organisms. The ubiquitous microbes, like bacteria, and the scavengers of the ecological systems comprise the part of the ecological and material flow system that cycles the rich chemical resources of dead organisms back into the system by making it available for the plants again. This is the part of the food chain that completes the important nutrient cycle. Aquatic ecologists usually refer to it as the detrital food chain.

Differing ecological communities emerged in differing environments, but even if bees and dolphins are not in the same ecological community, their respective communities do represent identifiable parts of the **planetary ecosystem,** which we all ultimately share.

While our local star powers most of the communities here, these ecological systems are only the biological aspects of the planetary system of energy flow. We have to consider that the ecosystems also depend largely upon the abiotic planetary processes (electromagnetic, geological, and atmospheric) which also cycle the materials and provide protection from cosmic threats, providing a safe home for the chemistry, biology, and ecology that is going on here.

The familiar nested pattern of a growing kind of entropically functional order is quite evident here on earth. Molecules, chemical reactions, autocatalytic reaction sets, prokaryotic cells, eucaryotic cells, organisms, societies, ecological communities, and planetary ecosystems all tell the same story of parts and wholes operating in concert to move and dissipate energy.

The much smaller scale realms in which we might look for nested patterns of order and organization also clearly continue the pattern. If we look at the Standard Model of Particle Physics, the collection of scientific findings regarding sub-atomic particles over the past 40 years or so, and simply ignore the details, the pattern is already visible.

The Standard Model allows us to see that particles called "**quarks**" are about as small as a structural, or mass-containing, chunk of matter can be observed with our current technology and knowledge. These have combined into composites called "**hadrons**". Hadrons include the very important **nucleons;** the protons and neutrons.

The protons and neutrons then combined (all in phenomena related to stars) to form the hundred or so naturally occurring types of **elemental nuclei**. These then combined further with neutralizing populations of electrons to form the **stable atoms**. Sub-atomic particles thus seem to clearly, but complexly, display the very same nested pattern observed elsewhere.

However, at about the level of organization represented by atoms, there is a proverbial fork in the developmental road of the universe. Atoms will continue be further organized, but up to this point their nucleus has been primarily held together by the short-range strong force. Now, they can be aggregated further, either by longer-range gravity, or by electromagnetism, which has already stabilized them with electrons.

The difference is profound. Whether an atom is further ordered, in a general Levolutionary sense, either by gravity into a dust grain, or by electromagnetism's chemical bonding into a water molecule or something more chemically complex, represents a branch in the development of functional order in the universe. For appropriate dramatic effect, I call these branches Axes of Functional Order. We can see a Particle Axis built mainly by the strong force, and a fork in which matter becomes further organized by either electromagnetism, which leads to chemistry and biology, or by gravity, which leads to astronomy.

Sidestepping the lack of a temporal sequence among the origins of **planets, stars, and galaxies**, it is reasonably apparent that the nested structure here, as elsewhere, is still generally a result of a sequence of combinations; of growing composites based on attractive force.

Everything named in bold here; virtually everything in the universe shows us the same pattern. It is a pattern in which entities, energy-consuming systems, form into sequential levels of organization. The process represents one in which each new type of entity is built from parts, and the parts are a set of the thermodynamically similar entities that already exist on the lower level that arose earlier.

This nested pattern of everything, from quarks to galaxies, and from atoms to ecosystems, is not an accident. Nor is it a speculation. It is a universally true observation that simply needs better scientific explanation than it has received. It is the purpose of this book to explain it fully as a result of Levolution. It is really an amazing story, and it is becoming clear, both to science in general, and to me personally, how and why it happened. It is a story that I think has never been told in its totality.

Eureka!

Strangely, it has turned out that working on the fifth mystery, the mystery of the cosmic order and the levels of organization, has led to an illumination of the other four mysteries. These are just five of the major mysteries that Levolution helps to solve. There are more.

I only set out to explain the process that builds the levels of organization, the nested structures of the universe, but like Archimedes, I am yelling "Eureka! I have found it." Unlike Archimedes, I will keep my clothes on, but Levolution is clearly the key to understanding the major current scientific mysteries of our universe.

I hope to convince you that the general idea I have stumbled upon yields something rather amazing in its implications. I will try to contain my excitement and focus upon being clear, but I want to emphasize that these discoveries seem very important to me.

Why Read this Book

This book is worth reading because it presents a very broad set of revisionist scientific perspectives that led me to a scientific epiphany. Most of these perspectives match well with observation and hang together logically in a way that reinforces their likely truth. I believe that this framework and body of ideas represent a very positive and unifying Paradigm Shift in physics and physical cosmology, and virtually every natural science.

What you will get are generally re-asserted and re-positioned principles, and/or newly proposed laws of nature, laws pertaining to energy and order. Several of these have been proposed recently (in the last 40 years) but are not exactly mainstream. They are supported, corroborated, and reiterated here in a cosmic context, and they find themselves among the Laws of Energy and Functional Order.

What seems most important here is the framework. The framework can hardly be called anything but an "ecological perspective". Ecological concepts, apparently foreign to normal physics, seem to weave the tale, but it is not just the bias of my training that causes this apparent truth.

Entropy's universal project relies upon dynamic systems to consume energy, degrade it, and dissipate it. In addition, each discrete system must do what we call "work," the use of energy, to create the entropically functional order that constitutes their system. Each discrete system has an environment to which it must remain adapted to transfer the most energy.

The situation of the universe as a whole, and with regard to energy, is much like the situation of an ecosystem, and as I have hinted, this same pattern can be found everywhere. I do not equate black holes with cosmic eggs, like Lee Smolin (1997), but I will point out that natural selection, evolution, and Levolution are all very much in evidence in the cosmos.

"Cosmic Ecology" is perhaps the name for this framework, and the type of narrative that seems to really work in explaining aspects of the cosmos that have been major mysteries. The big picture of the universe, the Large-Scale Structure, its components, as well as its underlying motivations, may all be gleaned from considerations of ecology directly. It also seems to have something to say about our own existence.

Levolutionary Cosmology includes a new explanation of the processes behind how the universe has developed from the unitary state of the Singularity, and a delineation of the processes at work, that resulted in the elements, the gravitational bodies, life, we humans, our cultures, and the planetary ecosystem.

While other authors have treated these subjects, my sense is that their tales fail to capture the very essence of all this, and it happens in one of two ways. Many authors are much more "mystical" in orientation than I am, and when they get into the real weirdness of the energetic truth, they immediately invoke their religious training and start talking about the spookiness of quantum physics, on one hand, or the spiritual influence of the divine creator on the other.

Still other authors have nearly stumbled upon the truth, but have been too reserved, cautious, or set in their ways to recognize a scientific revolution when they see it. Levolution seems like a new arrangement of scientific ideas that could and should result in a radical change in scientific perspectives about these subjects.

I know this all sounds like I am either crazy or over-reaching in the extreme, or both, but intellectual gems simply fell out of the work, in what I will call a very natural way. The Cosmic Ecology perspective has allowed me to stumble into a very important set of truths that has launched a landslide of new perspectives for me, and a tsunami of what seem like ramifications for science. I hope you enjoy this, the foundation of Levolutionary Cosmology.

SCIENTIFIC PROGRESS

Cosmology from First Principles

The Laws of Energy (the new thermodynamics) and The Laws of Functional Order (beginning with dissipative structures) are what I call the "First Principles" of the grand process of development creating the Cosmic Order in the universe, and simultaneously degrading, differentiating, and dissipating the energy.

The ten laws explained in Chapter 5, and listed in Appendix B, are sufficient to describe the general processes that provide direction, allow adaptive change, and promote the special patterns involved in both the origin and history of the universe. This includes the evolutionary and Levolutionary origins of all the material structures in nature, and the differentiation, through entropic degradation, that has created the many forms of energy.

Matters of Authority

Some of my associates think that various bad things will happen to me if I publish this work and demonstrate the audacity of taking liberties with the natural laws, especially those of energy and order. Thermodynamic Natural Selection and its related revolutionary ideas are set up to receive a major assault, even by fellow biologists.

There are many targets and opportunities to get lambasted by people smarter than I am, but the real problems will come if nobody reads it. There is no problem if someone reads it and disagrees with it. I will welcome any feedback, and hope that science carries on the dialogue.

Anyway, to give these good ideas a chance, despite both my lack of standing in academia and the non-corresponding magnificence of my ideas, here is what I would say in my own defense. It's good science.

The Role of Speculation in Science

Speculation in science, especially in physics, is sometimes frowned upon. There are many people talking about the sciences involved in nature and the universe, and there is a wide range of difference in their knowledge and orientations. For most purposes, it is best to adhere to the words of those with appropriate credentials.

However, there are intractable mysteries about the universe, and the current set of credentialed scientists are not exactly forthcoming with understandable and believable explanations.

From my isolated perch, I can see a number of reasons for this, but the biggest part of the problem is that there are few theoretical frameworks upon which to hang certain crucial aspects of knowledge. Other than the process of general cooling of the cosmos, the expansion in the inflationary universe, and the Standard Model of the particles there are none. Quantum theory offers no frameworks about causal reality at all. These are the only frameworks even available for organizing knowledge about the flow of energy, the growth of order, or the degradations of energy in the early universe. My aim is to fix that.

Paradigm construction by an outsider may be a good thing when the facts of the situation are rare, where placeholders may be needed to fill gaps in knowledge and assist in building an integrative framework for future knowledge.

Physics has its formalisms and its mathematical rigor, but on the way to such rigor and formalism someone has had to dream up or discover the ideas behind the numbers. Most current physicists minimize the importance of intuitive, rational, causal connections, but Einstein thought speculatively about light for years before working out his field equations.

In the ecology of islands, the principles of island biogeography include the notion that freedom from the normal predators, and isolation from the mainstream genetic stock are major factors leading to new species. Darwin's finches in the Galapagos are the most famous example. I think my own relative freedom and isolation from formal academia have led to some important and deep concepts. These warrant serious consideration by the mainstream.

The Role of a Paradigm in Science

A Paradigm is a framework for ideas that position them and link them together in a way contributes to a larger idea. If the links are mostly true, then altogether they will begin to yield another, larger truth.

When the knowledge that exists about a subject is unorganized by such a framework, the ideas are not linked, and fewer interactions among them means that they may not be tested together. Every linked idea in a working paradigm is automatically tested by the link itself.

The lack of a paradigmatic framework is the problem solved by the Levolution Paradigm. It is a paradigm composed of ideas that are based on a new, foreign, and ecological sort of logic. To put this point bluntly, everything has to eat something in the Levolution Paradigm. That is how energy flows.

A paradigm is thus a scaffolding or framework on which ideas, true or false, may be hung and connected. It says nothing really about the truth or falsehood of any individual idea within it. There could be a Paradigm based on pure fantasy in which the fantastic ideas link together in a single fantastic Paradigm. It would not easily connect up with the many other, more accepted ideas we share about reality, however, so it would be seen as generally false, but it would still be a paradigm. It is not particularly surprising then, that an entire Paradigm, a new and revolutionary framework, may be built from a set of laws, proposed laws, and observations.

Speculations are hypotheses, or trial facts, and suggestions regarding the truth. Their proof is something else, and when the assertions of the Levolution Paradigm are stated in testable form and experimentally tested, this will be a form of natural selection operating on the ideas. I have no problem with that.

My real point is that it is perfectly legal for me to assert new natural laws. The stodgy ones cannot sue me for thinking, and ideas have a life of their own. The assertion that the Levolution process is at the center of a new framework of ideas is surely in evidence here, but the mere existence of a scientific paradigm does not make it true.

Science is never done; it is always being tested. It is always subject to a kind of intellectual natural selection. Paradigm construction is a beneficial ordering of ideas that can reveal the best directions for research and experiment, can organize a set of theories or hypotheses, and is a tool in the hands of science. We are doing energy's work here.

The Material Subjects

There is no room for ambiguity in this important matter, so I must ensure that I carefully define the subjects of the universal processes I will lay out.

Many versions of "generalized systems" have been proposed over the years to serve as a construct in characterizing, in some kind of universal manner, the subjects of the order-building processes, or of processes related to general system behavior, feedback, and control.

Ludwig von Bertalanffy's "general system" was an early and important one that grew out of the study of biological ontogeny. Prigogine's "dissipative structure" is also one. Arthur Koestler, a social philosopher, coined the concept of the "holon" in characterizing his individual nodes of whole-part duality in the "holarchy" of living nature, as he put it. Stuart Kauffman's "autonomous agent" is another variation on the theme, emphasizing their ability to do a work cycle, and their autocatalytic properties as such may be captured in software algorithms as discrete, self-motivated, agents. "Complex adaptive systems" are yet another conception that John Holland coined.

My own conception of the generalized thermodynamic system as reflected by nearly all of the universe's real, entropy-producing entities, is the holosystem. It is my contention that the holosystem structure is equivalent in stature to a thermodynamic law, in the same way as the dissipative structure is. It is also my contention that it is the simplest and most fundamental way to capture the essential, but universal, characteristics of every system type on my list.

The holosystem was born out of a consideration of general system theory, dissipative structure theory, and Koestler's holarchy theory, and I think it is a very special structure.

Levolution

Levolution's work is the creation of holosystems, which are dissipative structures in terms of thermodynamics. They are also composites of parts, each of which are, somewhat surprisingly, also dissipative structures. They also usually exist within some identifiable boundary condition, certainly distinct from their environment, and so they exist as a discrete entity. These are all important traits.

New types of holosystems on new, higher levels of organization arise through events that are major innovations in the timeline of energy flow history. The process of their creation is Levolution. It has only happened about twenty-three times uniquely. Each new level of holosystem is a discrete aggregation of energy flows captured from the environment, and used to do work, they create and maintain order against the usual action of entropy, and finally to dissipate energy as entropy, on a wholly new and much larger scale, to its environment.

If you can look at something in the universe and want to classify it, one good way to start is to observe that thing's level of organization. You and I are multi-cellular biological organisms, for example. That statement puts the origin of our own holosystem type in a particular place in the flow of time, marked by an instance of Levolution.

There was indeed a time when a population or a colony of single-celled organisms began to differentiate under the influence of the holosystem pattern, an energy flow regime, and were integrated into a multicellular organism; a new kind of whole. It swam off its rock, found new environments, and evolved in each one. Eventually these things were writing science books about it all. Our general holosystem type thus evolved long ago, but we are still easily classified as multi-cellular organisms. The pattern left by Levolution is generally obvious, and generally known, but science has tended to leave it alone. I think that is because the process was missing.

There was a first quark, a first atom, and a first star. The first appearance of each of these innovations happens only once, of course, and when these events are marked on a timeline they show a kind of progress. Biologists have experienced the mistake of trying to make evolution's work, which is always adaptive differentiation, into a kind of progress. Evolution is related to adaptation, not to progress.

For Levolution, on the other hand, *progress* is the main phenomenon, and it is undeniable. Increasing the scale of entropically functional order is true progress. The innovations of encoded order and genetic reproduction were progress, and the organization of entire human cultures within ecological communities through the use of information-encoded, thought processes in brains, also represents progress. Energy has never flowed downward in potential any faster than it does today.

The number of levels of organization we might count, where most of the energy of the universe flows, is somewhat arbitrary, and I am trying to keep it simple, but there are about twenty-three clear levels of organization visible in the classification of all natural things. It is holosystems all the way down. Holosystems are a model thermodynamic structure featuring similar, discernable, and functional flow patterns for energy.

The energetic systems build on each other, with each new holosystem on each new level of organization being composed of parts that were formerly independent holosystems on one or more of the levels of organization that came before it. The holosystems are differentiated through natural selection, and become adapted into the complements, the parts, of a new whole, on a larger scale.

Each of the holosystem types listed above in Table 1 is composed of a population of parts that are represented primarily by the level just above it in the list. This creates the situation that downward in the list is upward in terms of holosystemic, entropically functional order.

Because of their creation, energy is degraded and dissipated faster in the universe, and this seems to be their, and our, primary reason for being. Note that of the twenty-three levels, seven of them are only known on earth. Thirty percent of the levels of organization in the universe have only been found on our planet, and I find this unexpected, amazing, and somewhat unsettling.

A Levolutionary Cosmology

In the final chapter of this book, after the new Laws of Functional Order are put in place, we will step back and look at the universe through the lens of the Levolution Paradigm. Once the holosystem model, the new and improved thermodynamics, and the dual processes of evolution and Levolution are understood, the real excitement begins.

Nature reveals a side that seems to have escaped most physicists. Fortunately, it did not escape Darwin, Prigogine or Swenson. With their help, the Levolution Paradigm applies universally, and it provides the essential clues about the earliest times, a new theory of gravity and *gravitational holosystems,* and the development of early particles as holosystems.

To this former ecologist, the main revelation is that all of these functionally ordered systems are hungry consumers of energy, specialized to capture, use, and degrade or dissipate some form of energy in some particular environment. Beginning with that premise, and some simple logic about the challenges of entropically dissipating the singularity, my ruminations have resulted in a series of strange, but logically connected insights. They are insights about photons, about the composite particles built from them, and the nature of the conceptual linkage between the differentiation of the various forms of energy and the formation of the earliest particles. They are insights about light, space, geometry, and gravity and the current mysteries of physics.

While certainly not proven, my speculations are at least logical, consistent, well-connected to each other, and they have a certain refreshing realism that is so utterly lacking in quantum descriptions of anything. Someday soon, I hope, they will be also be found to be disprovable, and backed up with mathematical rigor, both of which are important goals not achieved in this book. These ideas all sprang forth with virtually no effort, and certainly no experiments, but the logic of the Paradigm of Levolution pervades them.

Lee Smolin, a physical cosmologist who is clearly receptive to ideas from outside the box, outlined in his 2007 book, *The Trouble with*

Physics, a set of five of the major challenges facing physics at the cutting edge of current knowledge. I paraphrase them here as follows:

1. Combine the theory of General Relativity with Quantum Theory, resulting in a theory of Quantum gravity.
2. Add a dimension of realism and fix the foundations of Quantum Mechanics.
3. Build a Theory of Everything that explains the fundamental forces and particles.
4. Understand why the constants of the Standard Model particles have their values.
5. Explain Dark Energy and Dark Matter.

The Levolution Paradigm seems to shed a little new light on all of these.

Let's Get On With It

So, we have a novel set of revolutionary but well-founded theories here that calls our attention to a particular energy flow schematic, the holosystem model, and puts these discrete systems under the rule of a new thermodynamics, which now includes the Maximum Entropy Production Law and six Laws of Functional Order.

I have captured the crux of the Paradigm in the newly codified Laws of Functional Order, which begin with the definition of our subjects, which are dissipative structures and holosystems. These structures provide the fundamentally discrete, countable entities that happen to be required by quantum theory, but for us their importance is that discrete structures are required in order to describe the emergent operation of *thermodynamic natural selection.* Levolution's main mechanism is a new thermodynamic mode of natural selection, which is now one of the laws of functional order.

Thermodynamic Natural Selection is a profoundly important mechanism and it results in a process that, because it is not at all restricted to biology, we must now recognize as *Universal Thermodynamic Evolution.*

This is the process of adaptive change among dissipative structures and holosystems, and this universal mode of evolutionary change is also the mechanism behind *Levolution*.

These processes create and maintain the most important and successful pathways for energy flowing through the universe. Energy is flowing through a nested or *holarchic* pattern of holosystem types, and the result is everything in the universe. When I say everything, I really only mean the sub-atomic particles, atoms, molecules, the heavenly bodies, and all the entities studied as the subjects of chemistry, biology, psychology, sociology, and ecology. Outside of these areas, I haven't got a clue what is going on.

The holosystem model is explained here only about as well as Carnot's simple "heat engines" were in the mid-nineteenth century, and they are not that much different. Holosystems are simply the "dissipative structure" form, given material substance and physical reality by a group, or a population, of existing holosystem parts.

The details (which are the fodder of all the Natural Sciences) are not deeply explained in this very general and grand theory, but the generalities, because they are thermodynamic laws, will be central. Holosystems are real, fundamental, and universal, but only as a useful, functional template.

This situation hearkens back to Plato and his notion of "the ideal", and his theory of forms, which he called the "eide". The holosystem is like his master ideal, the "Eidos" from which the minor forms are all derived. While modern philosophers roundly condemn Plato's approach as incomprehensible and no longer useful, these folks have yet to hear of a dissipative structure, or a holosystem.

Maybe we should re-examine Plato's concept in this light of a universal thermodynamic, entropically functional model or form. Why would science be so perpetually interested in general systems, complex adaptive systems, holons, autonomous agents, and now holosystems, if we are not interested in an ideal, universal form of a functioning system?

The Levolution Paradigm applies well to particle physics, to gravitational systems, and to developing theory in autocatalytic chemistry. It applies well to biological, cultural, and ecological holosystems as well, and without any math, heavy lifting, or expensive tools.

Several previous authors have begun to see generally how the world has gotten where it is, and recognize the levels of organization developing to build it. Science is usually too partitioned into disciplines for this interdisciplinary view to be seen clearly, but on each particular level, the picture is becoming clearer. Autocatalysis among chemical reaction sets is a great new model of how chemical reactions can become spontaneously ordered (Kauffman, 1993). Recognition of the phenomenon of "endosymbiosis" among early prokaryotic cells (Margulis, 1970) is another great advance. The question has always been how such order is created.

My sense is that most scientists have "copped out" by accepting a notion of "self-organization" without a universal and workable theory behind it. The existing notions of so-called "self-organization" are not good science in my humble opinion. Self-organization always really involves natural selection from the outside, but because most theorists do not yet see the holosystem template carved in thermodynamic stone, they believe that the organizational pattern must come, somehow, from the inside. I believe that Levolution and the Laws of Functional Order will predictably replace "self-organization" and the related concept of "autopoesis" as the scientific explanation for the levels of organization in the universe.

Levolution, most simply put, is a process that builds a new level of organization from the action of a thermodynamic form of natural selection acting among one or more populations of holosystems.

The universal nature of thermodynamics, leads to a universal structural model (the dissipative structure), and Thermodynamic Natural Selection, operating opportunistically, but in accord with the entropy project of the universe as a whole, may create one by differentiating an already existing population.

Alternatively, natural selection may favor the mutualistic, co-dependent survival of a diverse set of already differentiated parts brought together or aggregated by other forces or phenomena. Either way, the former wholes literally evolve or adapt into the parts of a larger whole.

To see the entropically functionally ordered products of nature as holosystems is not so hard, mainly because you are one. I am confident that you will understand these energy-hungry patterns of energy flowing. All material things are built from one functional model; the holosystem.

The theory is general and based on thermodynamics, so it leaves the work of qualifying, extending, and explaining the details, but these general patterns are clearly visible throughout nature. Levolution has built the holosystems that are particles, atoms, gravitational systems, autocatalytic reaction sets, cells, organisms, cultures, and planetary ecosystems. There is not much room to dwell on any of these in this book, but the Levolution Paradigm turns out to be central to understanding the very architecture of the universe.

Heraclitus was right, although I do not know how he knew it without the benefit of science, and was so confident about it, in his time 2500 years ago. The universe that I can also see is truly one thing. It is becoming ordered and thermodynamically structured to move energy faster, and the details of the structural order are created by energy flowing and selecting the fastest pathways downward.

When these pathways become populations of discrete, dissipative holosystems, the population becomes subject to natural selection, to evolutionary adaptation, and they will get sequentially larger as a direct result of Levolution. All this is for the purpose of dissipating and degrading energy faster by achieving it on larger scales.

To be good in this universe is to be a good holosystem, and move energy as fast as possible consistent with survival, downward in potential. It is a relatively mundane sounding mission, but it is the one we have to accept. Let's hurry along, shall we.

CHAPTER 2

A NEW SCIENCE OF ORDER

Wisdom is one thing. It is to know the thought by which all things are steered through all things.

— Heraclitus

A BRIEF HISTORY OF ORDER

Order is not a simple concept, and to date, has not been recognized as an important part of physics. Philosophers, physicists, chemists, biologists, ecologists, and sociologists have all discussed it however. It is time that we gave it some deeper and more creative thought. Before we can understand the cosmic order, we will need to understand order generally. The following are just some highlights.

Heraclitus

While the ancient Greeks did not use a word exactly like "order" they did begin the process of understanding it in a scientific way. The concept is evident in Heraclitus's words above, and the way that he frames it is exactly the way it is. Wisdom is to know the order. The order however is itself just a thought. Order is a mere pattern in the way that energy (Eternal Fire) flows through all things. All things in our view here are the dissipative and holosystem structures of the universe, the seemingly material entities which are the subjects of Levolution.

Plato

Plato followed Heraclitus in his views about order, but he is famous for emphasizing that order is an ideal, or as Heraclitus called it, a thought. Plato's so-called Theory of Forms held sway for centuries and provided fodder for religious conceptions of divine immanence. His conception of an ideal world of forms and a separate material world essentially makes us, and all material things, into "craftsmen," who build the world, somewhat imperfectly, but based on the ideal forms which exist in another, more ethereal realm. According to Plato, the order observed in the material world comes from the influence of the ideal forms, which are intelligible to us.

Aristotle

Aristotle is a product of the same culture and was a student of Plato for a time. He broke away from the strictness of Plato's "two world's" picture but continued the same line of thought, and added some details. Aristotle's conceptions of order are a bit more down to earth, and are expanded into the Four Causes. These included Plato's Ideal Forms as the Formal Cause, which provides the "design" of ordered things, but he also enumerated three other Causes.

Material Cause is necessary to have the substance, or the parts, of whatever is being ordered. Things are not built from nothing. Material is a primitive concept, but in practicality, it is best to recognize the reality, and the order, represented already in materials.

Efficient Cause is necessary as the work, or the use of energy, in crafting the ordered subject. We are very familiar with the concept that a bill for repairs will include both parts and labor. Nothing happens without energy, or Efficient Cause.

Final Cause is the objective or goal that brings all the other Causes together to make the ordered subject. Goal-oriented activity is often apparent in cases of human-created order. Goal orientation is called "teleology". Where humans are not behind the goals, it was typically understood from the time of Aristotle, and through the Middle Ages, that nature has its own goals.

In Aristotle's paradigm of order, and in full view of his Four Causes, everything has its place in nature. Everything is arranged along a gradient of perfection that extends between Plato's Ideal realm, which has clearly become a conception of Heaven, and what may be a Zoroastrian (Persian) conception of Hell. The Divine is at the top and things become less perfect as one descends through what has become known as The Great Chain of Being. We humans are more or less in the middle, with angels above us, and animals (blooded and bloodless), and demons below. This was a very common view of the order of nature as it was understood up until the time of Newton.

Sir Isaac Newton

As Newton's Laws of Motion became distributed and understood, they managed to change the conception of order in nature. Instead of everything having a natural place in the Great Chain of Being, everything in nature moved, and kept moving, unless it was operated upon by an external force.

Motion, and the special case of rest, was a given, and outside energy was required to change directions, and to speed up or slow down. As applied to the heavens, Newton's ideas of motion led to an understanding that a common force called gravity operated similarly among apples as they fall to the ground, among artillery shells shot across the country side, the moon in its orbit around the earth, and the planets in their orbits around the sun. With Newton's few simple laws, all of nature seemed to be determinate.

Newton's paradigm of order then, is the deterministic view of the law-derived motion of projectiles. If the details of all the motion of all the particles and material objects were known, the future could even be predicted. The Greek, Democritus, had suggested that matter came in quanta that he called atoms. Atoms would make great projectiles to be discussing in the Newtonian context. The study of matter as possibly composed of atoms was not far behind Newton.

It was a hopeful order, but it was not quite correct. The biggest problem with it, however, was the challenge it laid at the feet of the atomists. How could they deal with an uncountable number of particles.

Carnot, Clausius, Kelvin, and Boltzmann

In an effort to understand the "particles in motion" as atoms, which were not known definitively in the nineteenth century, Ludwig Boltzmann wanted to understand the behavior of gases with the pre-supposition that a gas was composed of individual atoms in motion.

Boltzmann was impressed by Charles Darwin, who had discovered the Theory of Evolution by means of Natural Selection and had explained much of the biological order and the diversity visible among the plants and animals in nature. Boltzmann also wanted to explain order at a fundamental level, and sought an overarching law or relationship that would explain the behavior of gas particles in a way that resembled Darwin's discovery.

He did not see individual atoms of gas blinking differentially into and out of existence, but in another way he succeeded. Temperature and pressure were soon recognized to be emergent phenomena that, like natural selection, were exclusively a property of a population, or a plurality.

Pressure could not be attributed to an individual atom, but an individual atom could be understood in its kinetic behavior by measuring the combined effect of all the particles hitting the walls of the container. Boltzmann had discovered an emergent process, and it was profoundly related to the kinetic motion of gas particles at various temperatures. From this, Boltzmann utilized a statistical approach to quantify the relationship of pressure and volume to heat or temperature, and this led both to Boyles Law, and to the thermodynamic notion of quantifiable entropy.

Entropy was on everybody's mind because Rudolph Clausius had recently published his work and William Thompson (Lord Kelvin) had recently worked to understand the features of Sadi Carnot's heat engine. Lord Kelvin coined the term "thermo-dynamics" and included the regularity of "entropy" as set forth by Clausius.

Entropy had been discovered in one form by Lazare Carnot, and was developed further by his son, Sadi Carnot, who defined the work cycle and conceived of an ideal, reversible, heat engine. Rudolph Clausius called Sadi Carnot the most important researcher in the "theory of heat" and so Carnot may be the first name to associate with the discovery of the Entropy Law, but Clausius must be largely credited with setting forth most of its quantitative details.

Entropy in the context of heat engines says that some energy is always lost in the irreversible act of moving heat from one reservoir, through a working fluid, and to another reservoir, even in an idealized isolated thermodynamic system. There can be no perpetual motion machine because of this.

Entropy can be stated in about a dozen different ways, but some of these are not quite true. Order is definitely not always decreasing! The easiest way to think of entropy is as a tendency of energy to flow downhill in terms of potential. It goes away, somewhere, at every opportunity, and one can never get full use of any quantity of it. Boltzmann quantified the relationship between heat and entropy in terms of the kinetic energy of atoms and what we might call the statistics of distribution among atoms of gas.

Boltzmann's entropy relationship was statistical, but the only conception of order among the atoms or particles of gas for him was what I would call simply their "statistical proximity" to each other. As Boltzmann observed, there are many more distinct ways in which widely distributed particles might be arranged, than there are ways in which a set of proximally close particles might be.

Boltzmann's entropy was soon equated mentally with this type of order, or actually disorder, and the thermodynamic law was soon used to support a misconception that order must always decrease along with the potentials of energy. This seemed to correspond to the natural phenomena that occur among systems left to their own devices, but it only applies to systems left to their own devices, i.e. to artificially conceived, isolated systems.

The *statistics of spatial proximity* was then taken as a definition of something called order. Boltzmann may not be directly responsible for this error, but what most physicists take as a high degree of order is a pretty weak concept. It is nothing but particles in close proximity, a highly improbable, low-entropy, ordered state. This improbable state of what I will distinguish as thermal-kinetic order, however, is a state that has nothing to do with further increasing entropy production. Boltzmann's type of order is not entropically functional.

Needless to say, while physicists and chemists may have little problem with understanding order as fundamentally related to proximity, biologists had a little trouble with this notion. A biologist's view of order is more complex than this. A squabble ensued and has been raging ever since. It even rages now. The mysteries surrounding order are the ones that we most need to solve.

Albert Einstein

Before much else happened on this front however, Albert Einstein, a scientist, a patent clerk, a genius, by merely thinking deeply about the order of the universe came to some very novel and strange sounding conclusions. Einstein observed the constancy of the speed of light, pondered acceleration in the context of gravity, and used his own hypotheses. He used them to equate matter with energy, and to derive the effects these would have on the rest the universe. The results were his theories of special and general relativity.

If light's speed is constant, then things would appear differently depending upon the speed of observers. Time would tick at different rates, and space would have differing geometries in different gravitational regimes.

With virtually everything now relative to the speed of photons, science's view of order was seriously changed. Most humans don't have to understand it in detail because it only manifests at very large scales and speeds which we don't often encounter. In the calculations of astronomy, however, the fact of general relativity must be taken into account. The reality of variable spatial geometries must be faced by scientists. The order of nature is obviously not tied to our earthly or human experiences and intuition. Instead, nature obviously has a plan of its own.

David Bohm

David Bohm was a quantum physicist, and a contemporary of Einstein, who introduced notions of order that go beyond most treatments of the subject. He used his conceptions to attempt to build a better foundation for quantum theory. His unconventional ideas did not penetrate in mainstream physics despite his high standing, but I was surprised to find some resonances between the *implicate order* of David Bohm and the *functional order* of the Levolution Paradigm.

Bohm had more than one categorization scheme for order. In a simple way, he described a mere *descriptive order* in which a mathematical relationship might determine an orderly sequence. Contrast this with a *constitutive order,* which is an arrangement of actual parts. Bohm also recognized *generative order,* in which there is sequential, linear, and regular process of building some given order. Sequential creation of fractals by sequentially scaling and adding "similar differences" was his prime example of generative order, but life forms were also subject to this kind of generative order.

Bohm is best known for his conceptions of the contrasting *explicate* and *implicate* kinds of order. Explicate order is the visible, causal order that seems to provide scientific causes for events in the universe. Initial conditions are important, and physical laws operate to determine this kind of order. This leaves a causal chain which can be uncovered by science.

This reasonable-sounding, causal reality is underlain by Bohm's notion of the implicate order, which is a hidden order that is folded up, or *enfolded* in things, but exists in every part of a given whole. The implicate order may then unfold to create the more clearly causal explicate order. The implicate order is an "image of a whole" but emerges from each of the parts. Each part of a whole is hologram-like in that each part contains an image of the whole.

Implicate order is unfolded in a developmental way in Bohm's view by many aspects of the environment converging to affect the physical order of the subject. This influence is not clearly causal and determinate, but occurs due to what Bohm seems to say is a quantum phenomenon of some kind of emergence. The implicate order is contained, and hidden, in the parts, but emerges at the level of the whole.

In Bohm's view, the implicate order is an enfolded order, a hidden order that resides deep within the individual parts, but actually provides an image of the whole. The whole, once ordered, reveals the explicate order, the visible and causal order of the universe, but it's explicate causes were derived from an implicate order that was there all along.

Here is the salient point I want to draw from Bohm's rather obscure process of unfolding. For Bohm, the implicate order operates through the transfer of information in the quantum field. It is somehow this *information operation* that makes electrons behave as both waves and particles in the famous *double slit experiment*. Bohm's suggestion does not seem to have taken hold in mainstream physics. I was almost with him right up until the appearance of an information phenomenon. We will return to Bohm's ideas.

Lotka, Odum, and Prigogine

Some open systems, which allow the inward and outward flow of energy, enjoy a temporary period in which they suffer no such fate as increasing disorder. We can see, due to the persistent nagging of biologists, and the careful thinking of Ilya Prigogine, that in chemistry, biology, and ecology at least, there is a way that order can at least temporarily increase.

Prigogine described the phenomenon of dissipative structures in physical chemistry. Such structures are mere patterns in what may be characterized as flows of energy, but they are persistent and discrete patterns with the property that they align flows of energy creating the very patterns that define them. Indeed these are understood as "entities" or "structures" because one of the aspects of the energy patterns they adopt is that they become rather obviously distinct from their environment.

Dissipative structures, as discrete or distinct entities, seem to "capture" or take into themselves, energy from their environment. They actually also seem to "use" some of the energy to maintain the pattern that defines them.

Inevitably, of course, they also obey the Entropy Law, and so they degrade and dissipate energy to their environment. According to the Maximum Entropy Production Law, which is rather recent, the overall effect of their patterns of energy flow is always to maximize entropy production, which includes the expended energy of their work, and the "wasted" energy that they either cannot, or do not, use. These two outbound components of energy, if energy storage is ignored, must match up with the total inbound energy that is captured.

Dissipative structures are favored by thermodynamics because they speed up the flow of energy, and maximize total entropy production, but they do have limits. Dissipative structures are ubiquitous. All living things, including you and me, are dissipative structures.

We, and all dissipative structures, have limits as to the amount of energy we can flow through ourselves. The limits relate to the same laws of thermodynamics. We simply cannot maximize energy dissipation or entropy production if we are dead. Whenever we consider the Maximum Entropy Production Law, we must understand that system survival imposes this limitation. This is the natural order of things.

Complexity and Chaos

Chaos Theory represents a contribution to our understanding of order. Prigogine and Bohm both realized this as well. Randomness is an extreme lack of order. The order of a dissipative structure emerges out of chaos, spontaneously, with no apparent help, and if sufficient energy does not continue to flow, the order degrades back into chaos. The two labels are the ends of the spectrum of functional order. This implies a range of values if not a field.

Complexity is not as complex as one might think. It is usually acceptable to view it as the ratio of the number of parts per whole. Obviously as levels of organization build up, complexity is increasing, exponentially. Most of the things that have been written about complexity are matters of realizing the high degree of it in which we are immersed. I do not disagree.

Emergent Order

Emergence is the non-mysterious process by which a population of entities experiences some recognizable phenomenon, even though the individuals of the population, the parts, behave in ways that do not seem to cause the collective phenomenon experienced by the whole.

Emergence then, is fundamentally a "non-causal" communication from a lower to a higher level of order. It is not a communication of information, because it is not encoded. It is however a communication of energy's properties and its order, a translation between what individuals experience, and the experience of the whole that contains them.

Temperature is emergent from the kinetic motion of particles. Pressure is emergent from the collisions of the individual particles with the walls of a container. The individual particles do not have temperatures or pressures.

Natural selection is another emergent phenomenon. Individuals simply become unstable, come apart, fail, or die, but populations experience differential demise, or differential reproduction, of variant types and are experiencing the emergent phenomenon of natural selection.

David Bohm's implicate order is, I think, emergent. Emergence of a phenotype from a genotype is another way of talking about ontogeny and development. The cell emerges from the population of reacting molecules in the fertilized egg. The plurality or population aspect is defining for emergent phenomena.

Unfolding of the implicate order into the explicate order is not always the result of natural selection, as I will hold that it is in Levolution. Emergent or implicate order may occur from the operation of other natural laws and their emergent effects on a population.

What is important to note here, however, is that the emergent, implicate order caused by the indeterminate effects of natural laws operating in a population are not necessarily emergent from a randomized situation. Just because we cannot mentally follow each of the millions of acts of particles causing pressure in a jar, does not mean that the acts are random. The situation is similar to the difficulty that Heisenberg had with knowing where the electrons are in the shells of an atom. The task is too great for our minds, and so one could say it is our ignorance that causes us to see emergent properties instead. There is a thin line between randomness and confusion, but they are both matters of uncertainty.

Their quantized energy levels gave clues in that case, and have led to the notion of quantum mechanics. The particles causing pressure in a jar are also following the laws of thermodynamics.

Levolution is a matter of emergent order. It is not determinate, or deterministic, but is essentially the formation of a new and emergent order through the process of Thermodynamic Natural Selection.

The Order of Levolution

This story of the only universe we will ever know begins and ends with the phenomena of energy. Energy is literally everything, as the "Ever-living Fire" was for Heraclitus. The fact that makes this ultimate truth hard to see, in view of all the order, matter, and structure in the universe, is that the structures are real and durable patterns in the flow of energy. They are real and physical, but they are only generalized, schematic patterns, and the effects of the energy that runs through them.

The structures of matter are a side-effect of the universal project of energy, which is its own entropic dissipation and degradation at the maximum possible rate. I have explored to some degree the possibility that order-creation might be the real or primary project, and that energy's flow and persistent dissipation is simply the engine of all this order-creation. I have found that to be unlikely, but I will leave the question generally open. The two subprojects of nature, energy degradation and entropically functional order creation, are clearly inseparable, however.

We see evidence of nature's order all around us because functional order creation, where the function is to make energy degrade or dissipate faster, is essentially a "strategy" of energy in its downward, entropic flow toward a far-off entropic doom.

Order creation in the form of a dissipative structure, actually promotes increased energy flow and faster energy dissipation, and is a phenomenon that seems dependent upon, if not subservient to, the goal of increasing the entropy of the universe. As we shall see, it is also dependent upon the goal of doing so as rapidly as possible.

SIMPLIFIED THERMODYNAMICS

The subject of thermodynamics is sometimes confusing for lay people and physicists alike. I will help you sort out the modern and recently modified understanding of it, and offer some important recent advances. You may need to unlearn some things; especially about the nuance of order.

Eliminate from your thinking, for example, the notion that entropy is always associated with disorder. Many people go astray thinking along that line, and they will argue until they are blue in the face, but they are wrong. The war over order has raged since the 1850s, but we now have the needed perspective that will heal the rift. It will only happen, however, if certain people in high places will give it a chance.

It has been discovered that a certain pattern of energy flow represents a way to increase a new type of order, and this type increases overall energy flow and entropy production. Up until now, order has not really been studied as a scientific concept, and the distinctions I will make are perspectives to which the world seems almost oblivious.

Entropy, the thermodynamic law, is simply a statement of the direction in which energy spontaneously flows in nature. That direction is downward in terms of energy potential, which means that energy, over time, becomes more dispersed, dissipated, or otherwise degraded, to a weaker form. The law does not touch the subject of material order; it is about energy potential only.

There is a simultaneously entropic and cosmic, universal trend toward the elimination of gradients between differing energy levels or states. Two energy states that differ in their energy levels or potential will be met by entropy's directionality, and its tendency to obliterate the difference, erase the gradient, and converge the states on the same energy level; the lowest one obtainable.

Crucially, the Second Law of Thermodynamics says that entropy always increases, *not that order always decreases*. Entropy is just the directionality of energy's spontaneous flow. As we shall also see, an increase in entropy serves as the apparent "goal" of energy, and pursuit of that goal <u>can</u> result in increasing order. That is because entropically beneficial functional order increases the flow of energy. This entire book is about the increase in functional order that accompanies the cosmic increase in entropy.

<u>Entropy</u>

Rudolph Clausius discovered entropy, and gave it its Greek-derived name. Entropy literally means "transformative content". It is simply a characterization of a flow of energy with respect to a particular system. An energetic system, like Sadi Carnot's heat engine, conceptually makes a fork or a branch in the path of energy that flows through it. One fork or channel of energy goes into work, which is the use of energy that the engine was made to do. The other fork is entropy, and it simply exits the engine without contributing to its work. Whether some quantum of energy is entropy or work is a value judgment from the perspective of the given system. The name entropy refers to the fact that there is still some "transformative content" in the unused or wasted energy that exits the system. That energy may be "waste" to the referenced system, but it is not waste in general. It still has some transformative content. It is still energy, even though it did not even help a little bit to power the engine.

Regularities were discovered about the energy flowing in "open" systems. All naturally occurring thermodynamic systems are open to the flow of energy through them. We will only be concerned with open systems, generally because that is all there really is in nature. It is also true that most of the mathematical development of the science of entropy has been done on artificial (man-made) attempts to create "closed" thermodynamic systems. It began with Sadi Carnot, and his conception of the "heat engine", and the schematic diagrams of dissipative structures and holosystems are not much different.

If a flow of energy is being captured by, input, or fed into an open thermodynamic system, I usually call that the captured energy. If the energy can be used by the system, we will all call it "work" or use of energy. Work may be for any purpose of the system, internal or external. Work may be done internally to create and maintain the system's characteristics, patterns, or "order". A third flow of energy will always come out of an open system that is "entropy production" which may be degraded, dissipated, or simply unused energy that escapes or flows through. With regard to an open system, entropy is the outward flow of that part of the total captured energy that either cannot be used by the system, or has already been used by the system, and is thus degraded.

In discussing the energy of the universe as a whole, there are other perspectives to observe. Entropy is increasing in a sense that is not relative to a particular system as well. Energy is, in general, becoming more degraded and dissipated as it is used in the universe. Its energy is being degraded through uses made by energetic systems, and by discharges between potentials. As used in discussing the universe as a whole, entropy is not a relative evaluation, relative only to some system; energy as it flows is always headed downward in potential from any frame of reference. As Clausius stated the Second Law of Thermodynamics most simply, "The entropy of the universe tends to a maximum".

Entropic Teleology

As we saw, teleology is one of the significant philosophical products of the Greek philosopher, Aristotle, who lived several hundred years after Heraclitus. He famously enumerated his Four Causes of all things, which are *Material Cause* (material composition), *Formal Cause* (structure, form, or design), *Efficient Cause* (energy, work, labor), and *Final Cause* (objective, purpose, or goal).

This breakdown was very influential for many centuries, but the notion of Final Cause became associated, in the ways of the times, with the operation of the Divine. This development was intolerable to the scientific minded. In the Middle Ages, with the general birth of the scientific method in the sixteenth and seventeenth centuries, thinkers like Francis Bacon argued against use of the notion of Final Cause, also known as "teleology", in science, on the basis that mankind is not able to know the goals of the Divine.

This revolutionary step led to a more mechanistic view of science, and while the other three "causes" found homes in science, Aristotle's Final Cause did not. In general, science today does not consciously recognize teleology as an important construct, even when it is not tainted with Divine influence.

Entropy now poses a clear problem for this viewpoint because it is a universal process of energy in all of its forms, and so energy has an obvious, accepted, and scientifically understood goal or purpose, which is to degrade and dissipate. Entropy is the regularity we have discovered that causes energy to always spontaneously flow downward in potential. The goal of entropy is to reduce all gradients until the contents of the universe are uniformly distributed, motionless, and cold; the state that we imagine as a very distant entropic doom. I refer to this as the entropic goal, the entropic teleology, and to the process of getting energy to lower levels as the entropic project, the entropic mission, or the entropic drive of the universe.

I must generally assume that both Philosophy and Physics recognize this situation. I don't think it is my invention, but it is undoubtedly true. There is a Final Cause; there is teleology in nature. It is not necessary to invoke any Divine influence to recognize that entropy production is a very clear goal in the pattern of all flows of energy in nature, which we have long called the Second Law of Thermodynamics.

What this teleology does is allow the entropic goal of energy to be used as a valid reason, along with others, for things to happen. Entropy is a scientifically derived cause of all spontaneous energy-related events that really happen. Teleology is not necessarily appropriate in many areas of science, but entropy is an exception to the rule. Entropy is real, it is important, and it has a goal, even if it is a slightly depressing one.

This means that the concepts of "function" and even "purpose" are perfectly admissible with regard to entropy. Anything that promotes entropy production is "entropically beneficial," or "entropically functional," in the goal and purpose of moving energy to lower levels of potential. As we are coming to see, some specific structures in nature are very functional in this sense.

Maximum Entropy Production

With the aid of a thermodynamic principle that is even more recent than dissipative structures, but also related to entropy, it is possible now to see another important truth. Dissipative structures represent a way, a mechanism, to increase overall energy flow, and therefore entropy production, beyond the natural limit that exists when such order is lacking. Formation of dissipative structures is essentially a strategy or method to increase energy throughput and to *maximize entropy production* by creating "entropically functional" order.

Prigogine originally thought that such structures minimized entropy, because this is more consistent with earlier notions of increased order, but we can now see that this initial interpretation was wrong, and Prigogine and others observed that there is a distinction to be made between equilibrium thermodynamics and non-equilibrium thermodynamics. The rate of entropy production in the universe is increased by dissipative structures and they operate far from equilibrium.

While equilibrium is a "state attractor" related to minimal energy flow, dissipative structures actually represent another state attractor related to maximal energy flow, which is to say maximal entropy production, for a given morphodynamic situation. While equilibrium may be described as having a "basin of attraction" to its dynamics, our subject state or dissipative structure, has an inverted shape to its dynamics; a "dome of attraction". To get to faster flow, such a system may plot itself in many locations, but as energy continues to increase, its patterns of flow will converge on the dissipative structure pattern.

Increasing energy flow through a thermodynamic system, i.e. increasing energy capture and use, also increases entropy production. To avoid confusion I will almost never talk about any confusing concepts like the "stored energy" represented in a natural dissipative structure. Energy storage does not change the picture, even though it does add flexibility to it.

Entropy here is simply a characterization of one of two necessary energy output flows from any natural thermodynamic system. Such systems produce useful work, and such systems also produce a lot of spent, degraded, or dissipated energy that is no longer useful to it. Notably, when the work of such a system is done, the energy has been degraded and will then be evaluated as entropy.

Entropy may also be considered in the sense of the whole universe as a system. Energy becomes more degraded to forms that have less potential, are less energy-dense, or is more dissipated in space. The directional change in energy is always the same, both for a given system and for the universe as a whole.

Natural systems can use some of the energy they capture to create or maintain order, but a fraction of it is always unused, or unavailable and is transferred outside. Ignoring the temporary storage of energy by a natural system, every quantum of the energy consumed by such system will either be used, providing some kind of work output, or will be dissipated or degraded as entropy, providing the entropic output. An important thing to note is that when the work is all done, even that useful energy will have become entropy as well.

In reality, even the useful energy or work done by a system becomes degraded energy outside, and thus increases the entropy of the universe. This results in the notion that the total entropy of a dissipative structure is equal to the energy it consumes. Maximum Entropy

Production, a proposed thermodynamic law, is a whole lot like maximum energy throughput.

The spontaneous development of a dissipative structure's functional order represents an increase in entropy production to the maximum extent allowed by the constraints of its particular order. This is required by the relatively new, and very important, Maximum Entropy Production Principle. To maximize entropy production within constraints, a dissipative system will maximize its energy consumption, will use a fraction of this energy flow to create and maintain its special dissipative structure and functional order. The system will do whatever it does as useful work, but one thing is law given and assured; it will also release entropy to its environment at a faster rate than before it formed as a pattern.

While the idea that entropy production is always maximized is gaining attention now from physicists, it is really the contribution of Rod Swenson (1988), who is neither a physicist nor a chemist. That is one reason why you may not have heard of the principle. While the fact of it is very convincing, many physicists seem to suffer from various sociological aversions to ideas from outside their department. I am convinced that the principle is true and represents a new Fourth Law of Thermodynamics.

Regardless of what stodgy physicists think, energy always takes the steepest pathway downward in potential, within constraints. An important point is that one of those constraints for a dissipative structure is preservation of its own order and structure. Dissipative structure is essentially the same thing as entropically functional order, and the function that they promote is the maximization of the flow of energy downward in potential. The maximization of entropy production by a system is subject only to the available energy and the constraints of the system's continued stability.

Some scientists have noted the many constraints on energy flow, which characterize the very complex order represented by organic chemistry and biology, and have pointed to the fact that the constraints are the order in the flow. They then have used this as an argument against recognizing the principle that such systems maximize entropy production.

Here I think these thinkers must be missing the crucial fact that a dissipative structure would move no energy at all if it did not exist. Dissipative structures are, in a way, constrained to survive if they can. It is only within the bounds of their survival parameters that they can maximize entropy production and serve as a dissipative structure. They are doing the bidding of energy in their very existence.

There is an overarching principle involved among dissipative structures, and it is that the relevant constraints are their internal order, and even if energy flow is constrained by an orderly regime, the ordered way represents the fastest way downhill within the constraints that are necessary to keep the dissipative structure in existence. Dissipative structures have a limit to how much energy they can flow through themselves, or how much entropy they can produce, and stay within the constraints of survival as a dissipative system. I think that this may be a profound enhancement of the dissipative structure principle itself, and it ties in well with the points I will make about Thermodynamic Natural Selection.

Order is spontaneously created in the form of these discrete energy-moving systems, and it is created despite the entropy law, which works against another form of order. The process of a dissipative structure's order creation is essentially "driven" by the energy flowing downward in potential and finding the steepest pathways, while recognizing their own internal dynamic constraints.

THE ROLE OF ORDER IN THE UNIVERSE

Order's Origins

The Second Law of Thermodynamics is the Entropy Law. It says that energy is always headed downward in potential. Since the time of Ludwig Boltzmann in the nineteenth century, this has been taken by many to also say that something called "order" is also headed inexorably downward with the energy's potential. Order, in Boltzmann's case, and in common speech among physicists, because it has long been taken as accepted physics, is defined as a very improbable aggregation of the molecules of an ideal gas that might be contained in a sealed box. Disorder is recognized as the most probable situation, as there are many ways to be distributed throughout the box. There are fewer ways to be aggregated in one corner of the box in this logic. Most of the total patterns of arranging some finite number of molecules are patterns in which they are dispersed, dissipated, or spread out. There are more positions within the larger volume to consider, but there are the same number of molecules. This leads to the probability relation.

This stochastic conception of relative proximity is all there is to Boltzmann's type of order. Like quantum theory, it is a conception centered on statistical probability. It does not really say anything about the arrangement except relative proximity of the particles to each other. The spatial separation observed between molecules of an ideal gas when a sealed box of molecules is heated, is due to the increase in the kinetic energy of motion. The order, as it is commonly defined, decreases as the gas molecules disperse or dissipate due to increased kinetic energy input.

It has only recently, within the last 40 years, become known to science that flowing energy, in certain specified situations, is also capable of creating a type of pattern or form as it is heated. This is generally called *morphodynamics*, and examples are primarily demonstrated in fluid dynamics. The patterns form spontaneously, but by using additional available energy. Seemingly the pattern forms as an adaptation to an increase in the overall rate of energy flow, and this is the very purpose of energy's universal project. The special pattern in energy flow, or dissipative structure, that is formed moves energy downward in potential faster than it was moving before it formed. *Order can increase naturally through dissipative structure formation.*

The dissipative structure pattern was discovered by Belgian physical chemist, Ilya Prigogine, in the 1970s. The pattern is very special indeed. It provides the very definition of a kind of order that is different in character from Boltzmann's aggregated molecules.

These structures display patterns of a completely different type of order that is *functional* in moving energy faster. The dissipative structure is a generalized, schematic, and ideal form, an entropically functional, thermodynamically-defined pattern. With outside help and its own internal work to create order, it effectively, but not magically, creates and increases functional order. It is a universal, thermodynamic pattern that I have found to be exhibited throughout nature. It is the only means that nature seems to have to increase entropically functional order in the universe.

The academics are very conservative in this very fundamental area of science, and there is no need to upset them. Prigogine's discovery is settling in quietly as non-equilibrium thermodynamics. I simply observe here that "order" is not the simple thing that physics has long thought it was. There is also functional order, and the universe seems to be building it as its central occupation. The building is happening almost as what looks like a side effect of its entropic project to degrade energy.

Order comes in many types, and Boltzmann's type is *thermal-kinetic order*. This kind of order decreases as the energy level, temperature, or kinetic energy increases. Thermal-kinetic order is essentially meaningless as an "improbable concentration". Here the Entropy Law seems to mandate energy to decrease local order. This is how most people learned about order.

On the other hand, the type of order that we are interested in here is Prigogine's type, produced as the patterns or form of a dissipative structure. The order of a dissipative structure still relates to the energy level, but the energy of this special structure at the heart of universal and cosmic order now flows in patterns, has shapes to get into, functions to perform, and work to do. This is what I characterize as *entropically functional order*, which I sometimes shorten to *functional order*. When the conditions for this kind of order are present, and when constraints to it are absent, it may increase by a notch. This is just the opposite of thermal-kinetic order. It happens as energy flow increases.

This is the kind of order found in the open energetic systems of nature, not in the contrived "isolated systems" of physics laboratories. Even in the time since Prigogine's findings, it has become clear to many deep thinkers, but not all physicists apparently, that energy always acts to increase the entropy of the universe *as fast as possible within constraints*. This was a discovery by American psychologist, Rod Swenson. As a principle of energy flow, it produces a mechanism and a criterion for "energy pathway selection," and is called the law or principle of Maximum Entropy Production. Physicists for the most part seem to be trying to ignore it, but it is very important.

Following from two recently discovered thermodynamic principles, that I will here call the Maximum Entropy Production Law (MEPL) and the Dissipative Structure Law (DSL) we can suddenly follow the logic of energy.

Because dissipative structures are discrete entities, we suddenly also have the ultimate criterion and the potential for the profoundly important emergent process of natural selection. Because of the MEP Law, we have the selective parameter of the fastest pathways. Together, these lead directly to Thermodynamic Natural Selection among a population of holosystems, but that is not the end of it.

At the ultimate causal level, the selective parameter is always the same, and the responsive strategy of energy is always the same. By encouraging the fastest pathways, and letting them survive, Thermodynamic Natural Selection will build a new dissipative structure, if it can, on the next level up, from the population of, now part level, holosystems. It will evolve them into the parts of a new level of holosystem. These, after all, should always be found to be moving energy faster than anything else around, and so they make the best parts.

In ecology, it is the case that one is never sure what the criteria of natural selection actually are. There has never been a way to tell, even though we can measure fitness and count gene frequencies. It should help to have the thermodynamics as an underlying or implicate order.

It will guide new research aimed at connecting the two orders. Natural selection in biology is clearly not always based on the criteria of energy flow, except perhaps in the long term.

Natural selection may be several links away in the causal chain from a direct thermodynamic selection cause. This does not affect the theory, because thermodynamics is easily shown to ultimately trump virtually anything else. Whatever the causal chain may include, in the long run, the fastest movers of energy will have an advantage.

It helps to have the distinction between the various types of order, and it helps to know that there is a clear basis for both a Maximum Entropy Production Law and for a Thermodynamic Natural Selection Law. The former gives us an energy-pathway selection principle applicable to any pathways, even fluid ones, and the latter relating to populations of discrete and variable dissipative structures and holosystems.

The *discreteness* of the structures, as well as their *variations,* and their existence in *populations,* are all necessary to logically invoke the mechanism of natural selection as a matter of thermodynamics. These traits are all in evidence, however, and all backed up by thermodynamic principles, so we only need the right perspectives to prevail and Thermodynamic Natural Selection will have to be accepted.

The universe is pure energy, and it has a number of regularities that science has discovered. We call them natural laws. The laws about pure energy and its behavior are the Laws of Thermodynamics. The laws about "order" on the other hand, have not all been delineated, but we have started. New laws will be offered here, as codified in Chapter 5, as the proposed Laws of Functional Order.

The Entropic Project

While the idea of energy's descent to "entropic doom" or the "heat death" of the universe has been around to depress people for decades, few people really understand how deeply nature is involved in this Entropy project, and how it has been operational ever since the beginning of time.

The Second Law of Thermodynamics, the Entropy Law, mandates

this project of nature. The project is to move all the energy in the universe, the energy that first appeared as the Singularity, to ever lower, degraded, and dissipated energy states. We must now add *"as fast as possible, within constraints"* to that project description because we accept the Maximum Entropy Production Law.

The entropic project's fulfilment is the apparent objective of the universe; or at least it is the primary objective of the universe. It is not very inspiring, and it is not mystical, but it seems to be true.

It is ironically, a result of this strange objective, this goal, this "entropic teleology" that all the detectable functional order of the universe exists. It is the (entropically) functional order of the universe that leads to all matter, and all material structures and systems. About two dozen nested levels of organization among these things exist now. They did not exist at the beginning of time. The entities listed in Figure 1 are all dissipative structures.

Dissipative structures are not just convection cells in fluids anymore. They have built up in a succession as layers of nested composites called holosystems in the 13.7 billion years since the supposed beginning, when the universe was a point of pure energy called the Singularity.

The universe today is made up of discrete, individualized structures of many types, and they exist in large numbers. They are each energy-moving, functional composites, composed ultimately of pure energy, but they seem like "material" structures, which operate on various scales.

The structures of the universe, by combining into composites and joining together into ever-higher-order structures, have built up the scales and the levels of organization of what we call matter. This process of aggregation is Levolution. It is caused, as we shall see, by energy flowing and degrading, and the natural laws that govern the flow. The process is commonly, but mistakenly, called "self-organization," so I must explain why that is incorrect a bit later.

Today, science has discerned, or at least suspects, the existence of the following types of structures. We will be going into more detail in future chapters, but this is the Cosmic Order created by Levolution.

- The particles: photons, gluons, gravitons, up quarks, down quarks, leptons, nucleons, nuclei, stable atoms, and chemical molecules are all structures formed by various attractive forces, notably the strong force, gravity, and electromagnetism operating between opposing charges.
- Out there in the cosmos, we can start vaguely with dust, but we mean nuclei, atoms and molecules, which gravity begins to order, and we have the mass-ordered series: dust, planetesimals, moons, planets, stars, black holes, and galaxies. The objects in this series are all generally spherical or orbital, and in mass sequence, but they did not form in a temporal sequence, like virtually everything else.
- On the earth, and only on earth so far as we know, we have chemical reaction phenomena called "autocatalytic reaction sets" which will be explained later, and these aggregations of complex molecules and families of reactions are ordered into primitive cells, modern cells, and on to multi-cellular organisms, societies, cultures, and ecosystems. It happened in that sequence, although the planetary ecosystem was always there, and its population size is 1, so it is an outlier.
- At the very largest scales that observational astronomers can see, the governing order is not spherical, and we observe bubbles, called voids, and the clusters, and filaments. The filaments and clusters seem to me like they resulted from the effects of the inflating voids.

In considering the *aggregating principles* of these Axes of what I call the Holarchy of Nature, as adapted from Arthur Koestler, we can almost name them according to the form of energy involved.

Electromagnetism, the strong force, and gravity all conspired to produce the sub-atomic particles in the first series. Gravity is totally in control in the second series, and we will come to understand why time is not as definitive as space is in the sequence of this series. Electromagnetism returns to control in the ordering of chemistry, biology, ideational cultures, and ecosystems in the earth-bound series. At the largest scales we will have to understand dark energy and dark matter before we can really understand the biggest view of the Cosmic Order.

Electromagnetism is far more important than most of us know in everyday life. It is true that all of chemistry, biology, sociology, and ecology are all fundamentally electromagnetic, which is to say chemical, or electrochemical, or ideational in nature.

Chemistry and its phenomena, not to mention the energy of food, are ordered by bonds involving electrons. We even produce our ideas electrochemically in the brain. Human cultures and ecosystems are thus squarely on what we may call the Electromagnetic Axis of Functional Order, which I would note here is a straight continuation of the particle series, which began with electromagnetism.

The series of sub-atomic particles is complicated, but it is largely a series based on strong force attraction, which, with some help, has assembled the particles.

The tiniest particles with mass, the gravitons, and composites of them, simple quarks, are held together by gravity as well as the strong force. Electromagnetism is a minor player inside the nucleus, but is most heavily involved when electrons are attracted, and atoms and molecules become neutralized and stable by their incorporation.

I often just call this Levolutionary series the Particulate Axis of Functional Order, as a result of all this complexity, but we will get to the complex details eventually. Levolution has a simplifying effect on the particle zoo that has been uncovered by particle physics.

The Particulate Axis actually gives rise to the three others. The Electromagnetic Axis seems to be a direct continuation of it through the reactions of chemistry, but the Gravitational Axis of Order branches off where gravity appears. The fourth axis, the Dark Energy Axis, branches off there as well.

Dark Energy has its own very weak Axis of Order, and Dark Matter, which barely aggregates at all, resides on this Axis. While normal gravity has created most of the astronomical bodies out there, Dark Energy's structures do come into view at very large scales.

It all amounts to about 23 types of entropically and functionally ordered structures existing, in terms of their type, on three energy-defined Axes. Each of the Axes is a "Holarchic" series of nested structure types, but they are all connected by the flow of energy. They have formed sequentially over time, with the exception of gravity's ordering, which is a sequence based on distance and mass, more than time. These Axes of Functional Order, this Holarchy of Nature, and the stepwise descent of the Monodyne of Energy, and the processes that create them, are our subjects.

The Current Ordered State of the Universe

Levolution is about how the universe we find ourselves within has gotten to its interesting, highly ordered state. The state of the universe is highly ordered in terms of these structures now. It did not have planets, stars, atoms, molecules, cells, organisms or ecosystems in the beginning. Clearly, order is increasing.

The general direction of entropic energy flow in the universe is toward complete energy dispersal, but the constant drive to dissipate energy as fast as possible, just happens to sometimes increase functional order, even as the same energy might decrease Boltzmann's thermal-kinetic order. It seems time to straighten out the types of order.

The Entropy Law, the Second Law of Thermodynamics, provides a directionality to the energy of the universe. It tells us that entropy, the energy flowing in the direction that is toward reduced potential and toward less useful, less powerful forms, and toward more dispersed or dissipated states, is the direction the universe is going.

The problem is that there are several types of order, and energy flowing can lead to increases in one kind of order, even while it leads to decreases in another kind. Boltzmann encountered the kind that decreases as entropy production increases. Prigogine encountered the kind of order that is functional in increasing entropy production.

The twenty three levels of organization listed in Chapter 1 have generally formed sequentially over time. The universe is getting more order as time goes on. This fact introduces us to a scientific controversy, which has raged for 150 years, and still rages for some. This controversy became part of a widening gap of differences between physics and biology. It has been solved by the discoveries of Non-equilibrium Thermodynamics.

Biologists, dealing as we do with evolution on about six levels of organization, all from a chemical beginning, are quite aware of the fact that order has increased over time. Physicists, up until fairly recently thought that the directionality of the universe's order and organization was the same as its energy; always downward. Boltzmann seemed to show that order always spontaneously decreased, accompanied by energy's dissipation to lower levels of potential.

After all, left to their own devices, isolated structures will rot, rust, decay, and degrade down to their parts over time. Naturally-occurring structures are not left to their own devices, however. They are not isolated systems. Energy runs through most of them, and they are dynamic. Many are observed to actively work, or invest useful energy, against the degrading tendencies of entropy, to counter its destructive effects, and even grow larger, and more ordered over time. Particulate, biological, and gravitational entities all demonstrate this clearly, but it was once viewed as counter to the science of thermodynamics and physics.

The argument ran through science between the time of both Charles Darwin and William Thompson (aka Lord Kelvin) in the mid-nineteenth century where it started, and the time of Ilya Prigogine, in the late-twentieth century. The controversy, as I noted, has been solved, but most average people, and even many scientists, have not heard the news.

Its solution paints a new picture that science seems almost reluctant to see in its fullness. I aim to help in that regard. In fact, Levolution could be said to be the direct result of this controversy being solved. There are debts to pay.

Dissipative Structures

Ilya Prigogine's many years of work in physical chemistry resulted in his recognition of "dissipative structures" a type of thermodynamic model. One could properly say that this concept essentially "allows" the thermodynamic notion of spontaneous order-creation.

Be careful here, however, because "spontaneous" does not mean easy, magical, or self-accomplished. It only means energetically down-hill. This, more than anything, has led to confusion, and to the erroneous notion of "self-organization".

In nature, the kind of order we are going to be discussing, entropically functional order, can only be lawfully created through dissipative structure formation. There just is no other scientifically accepted process, in thermodynamics or elsewhere, for creating entropically functional order, spontaneously, among naturally occurring things.

Prigogine explained how the energy throughput resulting from work and entropy in certain ordered structures is greater than the energy flowing without the benefit of the pattern or structure, through the same region. The classic example, nearly always mentioned in this context, is the Bénard Cell, a set of hexagonal convection cells that form spontaneously and suddenly at a predictable temperature when heat is applied to the bottom of a dish of suitable fluid. The flows take on a characteristic, visible pattern or order, and the pattern actually increases the rate that heat moves upward through the dish. The convection cells are, in that entropic way, functional.

Dissipative structures, like swirls at drains, tornados and convection cells, have a visible pattern of flow and their order accommodates energy moving faster. Water goes down the drain faster with a swirl there, than it does without that pattern. Dissipative structures usually form as energy flow in the environment is increasing and reaches some threshold rate where the ordered pattern emerges, and energy flow rate plateaus.

At your bathtub drain, the flow of the individual adjacent water currents, if you could see them, will align in the same general direction and become ordered because this pattern allows the individual micro-flows to avoid random collisions, which slow the overall rate of the flow. At the macroscopic level, the flows spontaneously form a near-circular pattern, a whirlpool, or a swirl, and go down the drain at an accelerated rate, even though the circular order constrains their flow somewhat, as the center of the drain hole is now partially empty. Hold on to this concept. It will become more important.

Living things, as Prigogine himself observed, are also recognized dissipative structures. However, for us to discuss intelligently how order increases in the universe, we must go beyond the examples of fluid dynamics. Many more instances and examples of increasing functional order need to be given for all of the energy-form-related Axes. The cosmic, particulate, electric and chemical processes that build up the functional order must be enumerated, and outlined. This is difficult-sounding task is the one at hand.

The Science of Order

So, there is something very important called "functional order" and it is being created in the universe. The tale of its creation is unlike any other past or current cosmology in the sense that it is based on a completely unified view, and a theoretical foundation of natural laws, yet it still covers the entire range of the types of structures in the universe, each of which is functional in moving energy downhill. In some sense then, Levolutionary Cosmology is logically "tight" and could even be said to be simple. It is clearly a departure from normal physics.

Order is not usually discussed in relation to the sub-atomic particles, but casual references to gravitational order, which seems to be spherical or orbital, are common to hear. This new cosmology, however, portrays a universe that we not only are suddenly able to understand, but that we essentially do already understand. We just need a shift of paradigm, a new perspective.

Thermodynamics, in an expanded sense, and taken to include the offered Laws of Functional Order, supports the idea that energy's regularities and functional order are the keys we have needed to really understand the universe.

If we consider "order" as a scientific concept related to energy, systems, and evolution, and not to any other conceptions of the divine, we soon find that there are still various types of order in people's minds. Consider, for example, an "ordered population" of particles in a box.

One type of person would envision the order as the population of particles in the very unlikely and improbable state of being all concentrated in one corner of the box. Other people, in their mind's eye, would see the particles all connected by some kind of bonds and ordered that way into perhaps crystals. Others would look for evidence that the particles are really a colony of microscopic organisms, while others would see if they are in the pattern of a swirl at a drain in the box, or are perhaps convection cells. Order has several meanings because order comes in several types.

I will not dwell on chronological order, numerical order, or positional order, as these patterns are something else entirely and not directly related to our mission. I want to dwell on the type of order that has caused all the confusion in physics, biology, and thermodynamics.

Order needs to become a real scientific construct, despite the fact that it is always something like an arrangement or a pattern, and only physical in the sense that it is detectable and intelligible. The task of understanding the universe more fully lies in analyzing the concept of order, which science has long avoided. Let me start with this. We humans are very good at detecting order, and there are two relevant types of order that we encounter every day. One is Thermal-Kinetic, and the other is Functional from an entopic perspective.

Thermal-Kinetic Order

The Thermal-Kinetic type of order was studied and quantified by Ludwig Boltzmann. In concept, it is essentially "spatial togetherness" with regard to a population of particles or atoms. It decreases with the application of heat energy, because the motion of individual particles increases with the kinetic energy of heat. Particles spread out in nature due to the increased velocities and collisions among them, or the energy of heat.

Dissipation of energy, or the spreading out of heat, and even cooling and coming together in an orderly way, as in an ice crystal, are perfectly good ways to think about heat energy, and the Thermal-Kinetic kind of order. However, this Thermal-Kinetic way is <u>not</u> the way to think about order in general, because there is another important type of order to talk about.

Crystals, for example, are orderly arrangements of atoms or molecules in a base or rest state, characterized by insufficient kinetic energy to tear it apart. A crystal, however, just sits there as a structure or pattern. While thermodynamics is certainly involved in its formation and growth, a crystal is just a "spatially" bonded and ordered state of the atoms of the substance under the given conditions. The material order evident in a crystal may speed up conduction in some cases, but does not seem to contribute to the future flow of energy in any way that speeds that flow. The crystal is a matter of energetic constraints operating, but I do not see it as a matter of entropic functionality.

By contrast, a tree is an open system that has evolved and developed specifically to move and transform solar energy all the time, and its order or structure is actually functional in making energy flow faster. Crystal formation clearly represents a type of order, and it is worthy of further analysis in this context, but I will here dismiss it as a principle of spatial, positional, or thermal-kinetic order. It is a different phenomenon from our subject, which is entropically functional order.

<u>Functional Order</u>

Order is not merely the spatial togetherness of ensembles brought about by a lack of kinetic energy; it is also apparent as a phenomenon of patterns in energy flow. This type of order is associated with an acceleration of energy's flow. Functional order arranges the flows of energy within a region to allow energy to flow faster, to speed energy flow downward in energy potential. This, in fact, is the ultimate objective of the energy in the universe, and for this reason functional order is ubiquitous and important.

The contention here is that Prigogine, with his dissipative structures, has actually discovered a whole new *type of order* that has not received its due attention from science, because science has been preoccupied with thermal-kinetic order.

Functional order is the pattern of energy flows in all dissipative structures; the order of energy flowing in a functional pattern for the very purpose of fulfilling the mandate of the Entropy Law. This is the type of order that promotes the entropy project of the universe, and so I call it entropically functional order.

The region of space involved in, and enclosed by, a pattern of flowing energy in a dissipative structure may be generally called a "system," and energy would be observed to be flowing into the system from its environment, so this type of order concerns "open systems". The definition of the "open" system says that they are open to the passage of energy, both inward and outward, and this was how biologists made peace with physics for many years.

Thermal-kinetic order was studied in supposed "isolated systems" contrived by physicists, and so biology's open systems seemed almost to defy the Entropy Law by taking in some energy from outside and using it to repair the damage done by entropy. This kind of reasoning allowed biology's conception of order to "counter" the effects of entropy, but the modern way to talk about this subject is to invoke Prigogine's dissipative structure that represents the spontaneous creation of order.

The arrangement of flows is of interest, of course, because the patterns that develop in open systems actually promote the process of energy flowing. It's really quite a neat trick, and it represents a positive response to increasing energy. Energy flows faster in such systems, and this is how they got the name dissipative structures. Now we recognize all living things and many inanimate things as dissipative structures. Up until now, however, no theorist that I know has offered the hypothesis that all the particles and all the gravitational bodies, and all the chemical systems and life forms recognized in science as the dissipative structures that they are.

Dissipative structures always have a particular set of characteristic patterns of energy flow. These are always present, and able to be characterized as energy capture, energy use, and energy dissipation. The patterns are schematic patterns in flowing energy, and the energy of the medium flowing itself, but they are represented in the varying details of varying systems as well.

The system is defined as the region where the pattern exists, but the environment of the system is just outside of this, and it is where the captured energy must come from. The environment is also where the energy outputs of useful work and entropy dissipation must both go. Within this environment, the system must do whatever it is that it does to produce and maintain these flows of energy.

A simplifying fact or two will really help us simplify our discussion. To a dissipative structure, the outputs of energy will equal the inputs, ignoring any energy storage. The outputs, while properly distinguished as useful work to build or maintain a dissipative structure on one hand, and the degraded or dissipated energy, or entropy itself, on the other hand, will both contribute to the universe's race toward entropic doom.

Relativity is at work here. We must consider, not just the whole inertial framework as in gravity, but the holistic framework of the system-environment model, the part-whole model, that is operating in our labelling of work and entropy. Once energy flows through a system, and is no longer useful to the system, it will have become energy that is objectively degraded or dissipated. It is increased entropy from the perspective of the universe as a whole. It may still be useful to another system. Even the external work of a system that is used to build the local order of the next level results in degrading and dissipating energy to increase the entropy of the universe.

All this is true despite the fact that the system, the patterns, and the medium of both the system and its environment may all consist of the same material. In this sense, the entire intellectual enterprise here may be said to be "energy-centric". We will recognize systems and structures, but we will recognize that they are really pure energy flowing.

The regions in which patterns and structures exist are systems. Outside of these are environments. This is the *system-environment* perspective. It is important because it is the framework of ecological science, and we are going to generalize this framework.

The characteristics of the functional order-creating processes that have built the universe will be revealed herein. The processes of functional order creation did not have to be similar throughout the cosmos and on every level. We seem to have gotten lucky there, but I began looking for the general mechanism of such order creation many years ago. I must admit that I was surprised when I realized that dissipative structures provide a universal pattern that is followed throughout nature. I will explain this very unexpected finding by showing how each of the system types in the universe, be they gravitational, electromagnetic, or otherwise, is actually a dissipative structure.

The dissipative structure model is not a feature of living things alone, or of fluid dynamics alone. It is the structure of every naturally occurring energetic system in the universe. The dissipative structure represents the model for nearly every discrete "thing" in the universe.

It is a bombshell of a concept, mainly because it shows us a single ideal form, and then shows how every "thing" we know is, from a thermodynamic perspective, really in that form. It is pure Plato.

LEVOLUTIONARY PERSPECTIVES

The Paradox of an Ideal Thermodynamic Form

It might seem like enough to have the secrets of the universe, the thermodynamic process of cosmic functional order-building, land in your head. In fact, it did seem like enough, but it was hard to quit there. I had come to an understanding that there was a thermodynamically-defined structure toward which systems evolved, but in the back of my mind was a swirling paradox.

As an academic ecologist, I had been sure that natural selection is non-directional and purely adaptive. The whole crux of my education in ecology had told me that directionality in evolution was neither needed nor evident in nature. Adaptation to the environment seemed the entire objective of evolution; and achieving the objective, survival, was its own reward.

These teachings seemed well-founded, and were well-meant, but they are wrong. Ecology will need an update, and it does not seem like it will be a very minor one. Natural selection actually now has a target, both in ecology and elsewhere. It is thermodynamically directional, and the direction, after everything else, boils down to a targeted pattern that is functional in degrading and dissipating energy faster. Quantitatively faster overall energy flow is achieved by the energy flowing in the nested dissipative structure model. Technically, it is a maximization of entropy production.

The paradox is that we did not see the Ideal Form or target of natural selection because we did not have the Law of Maximum Entropy Production, and so we thought it was about the details of merely surviving within ecosystems.

Beneath ecological reality lies another reality of energetic concerns. The explicate chemical and biological order is underlain by the implicate thermodynamic order. In the game of nature, thermodynamics really trumps everything, and it provides the means for Thermodynamic Natural Selection to escape the shackles of biology and go on to conquer the universe.

Peacock feathers on the males of that species are as colorful as they are because of thermodynamic concerns. Even though this is the staple example and explanation for what Darwin himself discovered as *sexual selection*, a variant of natural selection involving mate selection, this mate selection process is a matter of survival as a genetic lineage or a species.

That apparent goal of survival is really thermodynamic survival, as a type of avian, multi-cellular, biological type of dissipative structure.

Entropy production is reason enough for pretty peacock feathers. Extinct species don't move any energy, and neither do dead peacocks.

We have always known that natural selection and evolution were about survival, but survival is now understandable as the thermodynamics of continued existence, continued energy flow.

The universe as a whole, and even the ecosystem as a whole, and each thing in it, as they are also thermodynamic wholes, has another priority. Natural selection is evolving the faster parts of faster wholes. Not only does it ensure adaptation to environments. It sculpts new levels of organization among the existing individuals and populations of the existing top level.

The criteria of the adjusting judgments of natural selection in the process of Levolution are really the criteria of what makes the best, and that means the "fastest energy moving, and survivable" dissipative structure out of the existing variant types of dissipative structures. The existing types, which arrived by the same process, will become the new parts of the new whole. They will change through natural selection. They will literally evolve into the new parts. This nested dissipative structure is a holosystem.

Holosystem parts will, creatively and innovatively, become well-adapted parts that aggregate and integrate to form a new composite; a thermodynamically defined composite that is a dissipative structure composed of parts that were formerly free-living dissipative structures themselves.

This all seems quite different from the normal notion of ecological "fitness" and simply adapting to an environment for survival's sake. Virtually all of the general laws of ecology, the study of biological systems and their relations to their environments, may now be seen to be underlain by Thermodynamic Law. H.T. Odum had tried to tell us this. Alfred Lotka, had as well. I will here give a nod and a wink to my former teachers who thought much the same, Ed Wiley and Donald Wohlschlag.

While the Levolution Paradigm brings unification and tries to heal the old wounds inflicted in the war over "order" that raged between the disciplines, it's tool is a blade that cuts both ways. Thermodynamic physics will invade and inform ecology, cosmology, and philosophy; but Thermodynamic Natural Selection will also invade and inform the ecology of wholes and parts, the cosmology of entropically functional order, and even the story of Levolution that underlies particle physics.

Underneath the surface of biology's sub-discipline of ecology, we find that the environment of each species is actually the inside of the system on the level of organization above it. Natural selection is revealed as the thermodynamic mechanism behind the Universal Evolutionary process that is continually evolving the parts. It is also behind the Levolution process that is making them fit into new, and larger, wholes. The wholes are holosystems. The ecosystem is a holosystem. Species are the functional parts of ecosystems. We knew that, but we have not really understood it.

Now, Levolution is simply the evolution of parts (always by natural selection), such that they compose a new type of system on a new level of organization. Evolution is simply adaptive change as needed to keep the parts all well adapted in the nested Holarchy of Nature. It is instructive to learn it like this. Evolution is adaptive change to a changing environment, but Levolution is adaptive to an opportunity, the opportunity of an entropy-promoting change to a faster energy transfer rate or regime.

Holism and Holosystems

The old concept of holism gets a bit more scientific as we delve into the formal subject of functional order. Holism seems to admonish us to view systems as wholes, not as collections of parts. I think it is a little more complicated than that. Levolution is restricted to a special class of dissipative structures. It is only about holosystems, dissipative structures that are composed of parts that are also dissipative structures.

What this recursive restriction does is to narrow the scope of our subject to only those dissipative structures that have clearly formed through the process of Levolution. As we shall see, this narrowing of scope to defined subjects does not at all narrow the implications of the theory, or even the scope of it. Almost every natural system that we study as science is such a nested thermodynamic system. I call these special, nested dissipative structures *holosystems,* and the holosystem is a thermodynamic ideal, a subclass of dissipative structure.

Holosystems and their Levolution

Holosystems are dissipative structures that are composed of dissipative structures as parts. See Figure 3 for a schematic of energy flow in their very special pattern. The list of holosystems in the first Chapter all have this pattern in common. There are other things in the universe, but they are not naturally occurring holosystems.

Figure 3—The Holosystem Structure

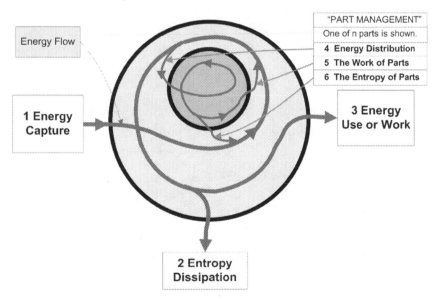

So, now I can say that this book is mostly about the Levolution of holosystems. The Levolution of dissipative structures into the successive levels of holosystems has built the universe and everything in it. I am not the first to see it, and I am not alone in seeing it, but the process that I call Levolution is more carefully dissected and more widely applied than it ever has been before.

Also, as housekeeping, I will refute the notion of "self-organization" which is holding Levolution's place in the ongoing discussions of science.

The Levolution Paradigm holds that the process of creating dissipative structures is the real driver behind <u>all</u> functional order creation, and it alone has built the seemingly "material" levels of organization we see in the universe. This has happened because functional order is a property, a characteristic, and strangely enough, a strategy of flowing energy to dissipate or degrade faster. Dissipative structure formation may simply be the only thing that works faster than conduction and the more primitive mechanisms of energy dissipation.

While energy spontaneously seems to order its own flows as it reaches some threshold energy throughput level, this does not really explain <u>how</u> functional order creation happens. "Spontaneity" in physics simply means it is a thermodynamically downhill process, not that it is simple or without cause.

Levolution, I contend, is what really explains how nature typically operates to create new instances of the more significant dissipative structures. Spontaneity and the often-associated notion of "self-organization" do not really explain anything.

While we are cleaning up and debunking, throw away "fluctuations" too, unless they have causes. Changes without causes cannot possibly advance science. This seems like a logical trick to make things seem statistical and open to quantum theoretical treatment.

Culminating Order

From the photon, which seems to be the universe's first dissipative structure, all the way to the planetary ecosystem of earth, and on another branching axis of order extending from the dust grain to the galaxy and beyond, something I will call functional order has been built, and it has organized energy flows into what we would call material structures.

This growth in scale, scope, and level of organization has been done by energy flowing in the only direction that it can; downward in potential. Dissipative structures help energy do that at every step in the creation of the levels of organization, also known by some as the Integrative Levels, of the universe.

Science has produced a number of thinkers who have come across and commented on the levels of organization, or somehow contributed to the science of them. I have surveyed these to some extent and find that I have both many commonalities, and many differences, with them.

There is a danger in being lumped in with some thinkers I do not always agree with, but there is no danger of real confusion with anything existing in the scientific literature in 2014. Levolution, in its thermodynamic essence, should not really be linked to the works of previous authors who have not tied in the laws of energy. It is fundamentally a work in thermodynamics.

Levolution is a new conception that started with my realization of the applicability of natural selection to economic enterprises. It grew until I could state the following findings about the Cosmic Order and its building in the universe.

The universal entropic project to degrade this universe's fixed allocation of energy as fast as possible involves two interesting and simultaneous subprojects. They are to:

1. **Differentiate, dissipate, and degrade energy.** The entropy project has differentiated the forms of energy from one Singularity into the many known forms, which include electromagnetism, space itself, dark energy, the strong force, the weak force, and gravity.
2. **Aggregate, organize, and build increasingly large composites** in the functional thermodynamic model of dissipative structures and holosystems. The entropy project has caused the development of a generally sequential series of composite

and entropically functional entities, built up from groups of integrated parts.

The composite structures build up the scale to handle more energy flow, and the use of energy by the composite structures degrades it sequentially to different forms, each with a lower energy potential. These two subprojects of the universe are conceptually detailed as the Theory of Levolution on the one hand, and the Levolutionary Theory of Everything on the other.

Entropy increases inexorably, but functional order does too. The degradation of energy actually powers the process of making the order. One question about this emerges quickly: Which of these is more important or primary? Certainly, order-building might, for us, take the moral high ground, but energy's entropic project actually seems to be the most fundamental goal of the universe. Putting a thumb on the scale so that the answer comes out to be cosmic order creation, the question then becomes whether energy's degradation could be simply the engine necessary to produce the universe's amazing order.

Because of this dual action of energy flowing in the universe, this Paradigm leads surprisingly to both an explanation of the cosmic functional order, and a Theory of Everything.

In physics, a Theory of Everything has been a long-standing objective, as it would explain how all the forms of energy are related and unified. The goal of explaining the structures of the universe in terms of the creation of functional order was my only goal in the beginning, but because this effort caused it all to fall into my lap, a plausible Theory of Everything has emerged and will be offered soon in another book.

CHAPTER 3

LEVOLUTION IN SCIENTIFIC CONTEXT

Though this Logos is true ever-
more, yet men are as unable to
understand it when they hear it
for the first time as before they
have heard it at all.

– Heraclitus

LEVOLUTION IN A NUTSHELL

Levolution is a way to understand the process that has caused the multiple levels of organization that are obvious among the things in the universe. Levolution's process of building new types of systems on new levels of order is what has caused the observed nested sequence in all the material structures of the universe. What then caused, and still causes, the process of Levolution to occur?

The process of Levolution is not programmed into matter as in archaic "vitalism" and it is not correctly described as either "self-organization" or "autopoesis". It is not a matter of reaching a "critical state" through random fluctuations.

One good way of viewing the causes of Levolution is to follow Aristotle, and consider his four causes of things. These were material, efficient, formal, and final causes. We will modify these somewhat to discuss the *efficient driver*, the *formal template*, the *material enforcer*, and the *purposeful teleology*. These are the causes of Levolution.

Simply put, there is an energetic *driver* of action and a *formal template*, or specification, for the action. There is also a process to *enforce* (materially in the real world) the specification of the formal template to maximize entropy production. Finally, there is the *teleology of entropy* to provide the purpose for the action, and the Levolving system type.

The Driver and the Template

The driver of Levolution is energy itself, considered as combined with its apparent final cause, teleology, or purpose in the universe, which is to degrade and dissipate to the lowest possible potential as fast as possible.

This is to say that the Entropy Law dictates the direction in which energy must go, but the maximum rate requirement is dictated by the new Maximum Entropy Production Law. This law comes from our observation of energy's selection of the fastest energy pathways that are both available and consistent with maintenance of the template pattern.

Given energy's dictate to flow where and when it can flow downward the fastest, the only other thing needed is the ordering of energy's flowing, or arrival at the <u>one pattern</u> in which it might flow to achieve an increase, a maximization, of the rate of flow.

The special thermodynamic form, the dissipative structure, is a pattern that, when energy actually flows in accord with it, allows the rate at which energy flows to be faster than it was before. The energy also attains a pattern that we clearly recognize as a discrete entity or a dissipative system.

It certainly looks to me like energy came to this universe with its own regularities, which we understand as its Laws. If we expand these from classical thermodynamics to include Maximum Entropy Production as a new law, we get the efficient driver and the final cause. If we then add the new Laws of Functional Order, which includes the Dissipative Structure Principle as the first one, we get the template too.

We can then say that the template of an entropically functional order, a thermodynamic structure, an Ideal Form for moving energy, came right along with energy and its tendency to dissipate. These are grand assertions, but that is only because of where they are. They are at the thermodynamically fundamental base of the logic for why there is material structure in the universe.

It is a statement of the universe's *entropic project*; to reduce energy potentials and gradients as fast as possible. The increase in scale and scope that accompanies each act of Levolution translates into faster and more modularized energy flow, downward in terms of its potential, and that is the apparent objective of all the energy in the universe.

It is important to view the cause of Levolution then, as an event driven by, and consistent with, the goal of energy degradation and dissipation.

The formal cause, or the design element, of Levolution is the functional template that nature always uses to make energy flow faster through a population of systems. It is through the manipulation of the systems into a special set of universal, complementary functions; all related to energy flowing as a single dissipative structure. The experience of a population of systems in the real world changes them, enforces the template, evolves them, and turns them into a differentiated set of integrated and complementary parts.

The science of manipulating populations of systems into differentiated, but integrated, parts of some larger whole already exists. It was the subject of my academic career and it is called *ecology*. The mechanism behind all the phenomena that add up to Levolution is nothing other than natural selection acting to evolve what were once free-living wholes, into whatever are needed as the integrated and dependent parts of a larger whole.

Bigger is better when it comes to making energy flow faster, so we have *growth in size* as a common result of Levolution, while with evolution, that general tendency is not quite as apparent. Even though it did not work out well for dinosaurs, they are an exception to the general thermodynamic rule. Developmental growth in biology is driven by a similar thermodynamics, but the process of growth there is programmed. Considerations of size among evolving animals is a fascinating subject, but one that I will ignore for now.

Consider rather the Levolution to multi-cellularity that happened earlier in evolutionary history. That represented a large size increase for organic life. If the environment allows an event of Levolution, if constraints allow, it is likely to happen, because the process is thermo-dynamically downhill, or spontaneous. It results in faster energy flow.

The various constraints on naturally occurring, energy-using, working systems, and the details of how they produce entropically functional order, are numerous and complex. The process of Levolution has only happened about 23 unique times in 14 billion years. To me, this seems to say that while the process of Levolution is thermodynam-ically spontaneous, it is still not very easy to accomplish and maintain. Constraints and the contingent details of the environments of the sub-ject systems are important limitations. Levolution is not a picture of determinism, but is one of myriad possibilities, a directional driver of change, and an entropic goal or teleology.

Thermodynamic Natural Selection

Thermodynamic Natural Selection refers to the mechanism under-lying Universal Evolution, and it is a process created by the union of two modern thermodynamic principles related to non-equilibrium thermodynamics:

(1) the *Maximum Entropy Production Law (MEPL)*, which says that energy always selects and allocates itself to the steepest, or fastest, downward pathways available; "downward" referring to the direction in terms of energy's potential, and

(2) the *Dissipative Structure Law (DSL)*, which says that discrete entities or systems may spontaneously form as a pattern, or may be or-ganized and structured by flows of energy. These very special structures have the property of increasing the rate of energy flow through them-selves by virtue of their entropically functional ordering, or arranging, of the flow pathways.

These two very important principles, the first from a

neuropsychologist, and the second from a Nobel Laureate in chemistry, are emphasized herein to the maximum extent possible. They are then logically combined to create the enforcing process evident in both evolution and Levolution, the new law of Thermodynamic Natural Selection.

Both are necessary because natural selection, as that mechanism is technically understood from the well-developed science of ecology, refers <u>exclusively</u> to an emergent, population-profile-changing process affecting populations of varying, discrete entities.

The MEPL only defines the behavior of energy flow pathways, which are not countable and discrete, but the DSL provides discrete entities *as such pathways,* and thus unlocks the potential for variable populations of them. These concepts, intellectually and logically, pave the way for Thermodynamic Natural Selection to be the main agent of change in every type of dissipative structure or holosystem. By "change" here, I simply mean the modification of the characteristic feature profile of a population of systems; the expressed mainstream type of the population.

Universal Evolution by Natural Selection

I have explained the *efficient driver,* which is energy, the *final cause or purpose* of entropy, the *template,* which is the dissipative structure, and mentioned that natural selection is *the material enforcer,* involved in Levolution. The material cause is where the ideal and the real meet face to face. The *material enforcer* component of the cause of Levolution is the law and the fact of Thermodynamic Natural Selection. This profoundly important process creates a universal mechanism of adaptive change, which can now be applied logically to populations of any type of dissipative structure in the universe.

Universal Evolution, as the words are meant here, does not mean merely "change over time"; it means Universal Evolution by means of Thermodynamic Natural Selection. Physicists will have to come to grips with this reality, but they will predictably have some difficulty because they are habituated to use the word "evolution" with the simpler and unexplained meaning of mere change over time.

Evolutionary change does not really require reproduction, mutations, generations, or fixed lifetimes. These happen among the living things, and also happen among some gravitational bodies. Evolution by Thermodynamic Natural Selection even happens among the subatomic particles. Natural selection can cause change in a population of things simply by subtracting certain variants and changing the mix, or the resulting profile of their features.

The mechanism of natural selection is completely independent of the type of things it adapts. It applies as well to a game of checkers as it does to wars among cultures, and drug-resistant microbes. It is grand in its scope, fundamentally thermodynamic in its nature, and you might say it is a force for good in the world, as long as good means "faster" in terms of energy flow and dissipation.

Natural selection operated early in the history of the cosmos, and it does so today. It operates in accord with Newton's Laws out there in space. It operates in accord with nature's ecological laws on earth. It operates as you shop and make choices, as you edit, and as you read and process the written word. This simple mechanism is nature's primary means of creating adaptive change. It is used everywhere you look, once you open yourself to the prospect. When the truth of it really sinks in, you may wonder how the universe could be any other way.

I follow Williams and DaSilva (1996) in observing that natural selection is almost synonymous with the notion of *"constraints"* used by physicists and chemists. Both concepts determine what can or will exist in the future.

Constraints on systems may be either internal aspects of their required functional order, or externalities that limit action and govern survival, but the physical notion of constraint is usually a hollow concept, a placeholder. In the informal talk of physicists and chemists, constraints have lacked specific meaning, and lacked a theory, but it is all visible within the Levolution Paradigm. Internal order constraints are what enforce the order operating in the system. External constraints are the aspects of the environment that cause differential stability and survival. Both notions of constraints may now be viewed as aspects of Thermodynamic Natural Selection.

Constraints operate as a matter of Thermodynamic Natural Selection, and scientists need to drop their biases and confess to it. This is an interdisciplinary merger of theories that I can predict will broaden the horizons of future scientists. There are many operating constraints in the complex systems, or autocatalytic reaction sets, of living chemistry.

These constraints, especially those that seem to meter, or slow, the energy flow in biochemical systems, have been part of Stuart Kauffman's reasoning for suspecting that something other than natural selection is at work in his theory of the origins of life and its order (1995). After much thought, I have come around to the idea that he is right about this. There is something else operating besides natural selection, but it is not self-organization, as he suggests. It is the ideal template of the dissipative structure, which is always the target of Thermodynamic Natural Selection.

Levolution as the Primary Mechanism of Order Creation

The fact of Universal Evolution by Thermodynamic Natural Selection leads directly to Levolution, because it provides for adaptive change of wholes into the parts of a larger whole. There is, however, an aspect of natural selection that has not been recognized to date. It is the fact that natural selection, as a thermodynamic principle, also has a thermodynamic target.

The targets or purposeful endpoints of natural selection are always adaptations toward the needed parts that would complete a complementary and integrated set of energy flow pathways that represent the functional pattern of a dissipative or holosystem structure.

Usually these target forms are achieved by evolving the existing, highest-level systems into the parts of the next higher one. If this were not beneficial to survival, it would not happen. If this were energetically costly, it may not happen. It seems to be what almost exclusively happens. The universal project of energy is to attain lower energy potentials, by flowing downward at the fastest possible rate, by reducing energy gradients or potentials as fast as possible.

This is typically accomplished in nature, not by heat conduction, and not by simple dissipative structures, or convection cells, but by holosystems. Some new words are necessary to succinctly talk about Levolution.

Holosystems are dissipative structures that are also built from a population of dissipative structures on the level of organization just below it. Below it here means in the mass-based rankings of entropically functional order. When increasing entropically functional order is graphed as an ascent, then "up" in this ranking has the understandable meaning of more ordered, higher in order, or on a higher level of organization.

The ranking list of holosystems (See Figure 1) is actually another kind of important, thermodynamically-inspired, structure. It is a *Holarchic Structure,* sequences of nested parts and wholes. The parts and wholes are all holosystems, as these definitions are constructed, and the overall Holarchic Structure is the conceptual structure of the types of things in the universe. Arthur Koestler called it the "Holarchy of Living Nature" and thus coined the term.

The holosystem structure, the subject of Chapter 4, is the target structure of Thermodynamic Natural Selection. New holosystems originate through the Universal Evolution of lower-level dissipative structures into the evolved and differentiated parts of a larger dissipative structure, a whole, a holosystem. Holosystems and Levolution are paired phenomena, and they replace the long-standing fantasy of self-organization.

THE CONTEXT OF OTHER THEORIES

Ideas do not occur in a vacuum. They have their own ecology within individual minds and cultural institutions, and I might add that the logic above lays bare many mechanisms at work with ideas. My ideas initially came to me as a result of exploring the relationships between ecology and the cultural-ecological science that we call economics.

Similarities among the levels of organization, and the existence of the levels of organization themselves, are what spurred me. But as time passed, I realized that what I considered then as novel concepts, were indeed being addressed in science in some way. I have read the works of many authors who were on the trail that leads to the Levolution Paradigm, but there are many cases where I had important differences.

In this Section I will briefly touch upon some of the more important sources of what I would view as error in the incumbent, or current, paradigms. This is not intended as any full-fledged argument against them, but merely a matter of pointing them out. I certainly do not indict science for its progressing, or any of the involved workers for working, even if it produces occasional errors. Errors are part of the scientific process, and science has both the objective and the means to correct them.

These statements of my personal positions on issues are also not knee-jerk reactions against competing ideas. Being old, I have learned that failure to appreciate something is often simply due to a failure to understand it. But here, unlike in most of the book, I am saying that I understand these issues and have reached a point of departure from them nonetheless.

The Trouble With Emergence

It seems to have all begun with Chaos Theory, which provided a view of a kind of order being spontaneously produced in chaotic regimes, even when they were not produced by artifice. Ilya Prigogine's dissipative structures were an example. Oldershaw's fractal universe was another. Coastlines even appear as self-similar fractals, with scale-invariant properties, for another example.

There is a surprising tendency of the universe toward some kind of order that exists right under our noses. I was hooked by the subject. I did not discover David Bohm's conception of *implicate order* until much later, as Prigogine's *spontaneous order* slowly sunk in, and I realized its profound importance.

Chaos Theory and spontaneous order morphed into Complexity Theory, and computers were applied to help explore the algorithms that might be at work to produce instances of visual, pixelated order among computational entities, usually referred to as "agents". The work of John Holland and others at the Santa Fe Institute were important here. Some interesting and general conceptions emerged from this work, and one of them is the concept of "emergence" itself.

Emergence was once a candidate for the title of this book, and I was relatively certain that I was about to be scooped when I saw such titles as Harold Morowitz's 2004 book, *The Emergence of Everything*. Emergence as a physical principle, however, was not new to me as a population ecologist, and was not sufficiently grandiose. Emergence is simply the mental "math" of thinking in terms of populations. It subsumes those individual agents whose behavior may, or may not, create higher-order patterns in the population. It is a change of reference frame, much like the mental gymnastics of general relativity. Levolution is an emergence, but emergence is not necessarily Levolution.

Emergence is just a population effect that is caused by individual actions of members of a population, but one that results in a pattern only at the population level. It is just like natural selection, which does not occur to individuals, but is a pattern that can only be seen at the population level.

Pressure is another common emergent property. Individual atoms or molecules collide with the walls of a container, but only the container as a whole has the parameter of pressure. To those who use these concepts routinely, emergence is an intuitive kind of math, done as a short-hand, and in the head. People like Lotka and Boltzmann have reduced it to real math.

Levolution is not mere emergence, even though it clearly is related to it. Levolution is the formation, by Thermodynamic Natural Selection, Universal Evolution, and Levolution, of functional order in the form of the holosystem types. Natural selection is itself emergent.

The importance of Levolution is not in the mathematics or the algorithms of emergence. It is in the nature of the actual functional order that is emerging when a set of actors or agents *differentiate* and *integrate* to produce a holosystem. The importance of Levolution is in the holosystem structure and in the functional order that holosystems build into nature.

Algorithms will be written along these lines soon enough, but they will need to capture the thermodynamics of holosystems and the Laws of functional order. The biggest difficulty, and possibly the show stopper, will predictably be in modelling the thermodynamic reality of natural selection acting naturally. In other words, it is the sometimes surprising, real, physical challenges to holosystem performance and survival that will need to be captured in software. This does not seem like an easy task, but people connected to computers are very smart.

The Trouble with Spontaneous Order

Ilya Prigogine's work in the 1960s and 1970s culminated in his Nobel Prize in Chemistry in 1977. His most important breakthrough was the logic behind a type of thermodynamic structural model for energy-moving systems (Prigogine, 1980) that results in what I here recognize as a new and *entropically functional* kind of order, the dissipative structure.

While he worked out the logic resulting in the creation of such morphodynamic systems, his main contribution seems to be the simple allowance of them; the very possibility of them in nature. Biologists had already recognized them as some kind of vague exception to the rule. It took a physical chemist to sort it out in a proper language and style.

Dissipative structures create a particular kind of order. Order comes in various types, and if it has been treated scientifically at all, it has been treated blindly. The thermodynamics of entropy is very much involved in both thermal-kinetic order, which Ludwig Boltzmann was able to deal with statistically, and with entropically functional order, which is completely different.

Boltzmann's order is simply statistically improbable arrangements of particles in random, heat-induced motion; the notion of particles all residing in one corner of a box. The order that Prigogine uncovered and legitimized, however, was entropically functional. It actually helps in moving energy downward faster than before. That is why I call it *functional order.*

Not only are dissipative structures possible, according to Prigogine, they form spontaneously, where "spontaneous" means thermodynamically "downhill". In particular, these structures are not obstructed by, but are driven by, the Entropy Law.

The allowance of dissipative structures is a very important contribution that solved the problem that biologists have had with the Entropy Law since the time of Lord Kelvin, the inventor of what he called "thermo-dynamics". That problem, of course, is the order that is so obviously produced in evolution by natural selection. Evolution flies in the face of the raw Entropy Law because that Law was thought to always mandate increasing disorder, not increasing order.

The conflict may be resolved by the recognition of two kinds of order; thermal-kinetic order, which always decreases and functional order, which may increase through the spontaneous formation of dissipative structures and holosystems.

The Trouble With Self-Organization

Modern science's approach to this subject began with W. R. Ashby in 1947. Ashby's conception of self-organization has merit, and I can recognize that he was seeing a dimension of Levolution. His conception utilized the construct of a state *attractor*, a basin of attraction in dynamics, in which systems are spontaneously directed toward the physical and thermodynamic ideal state called Equilibrium. Subsystems of a system, in his view, had adapted to a kind of equilibrium state in the environment formed by all of the other subsystems.

Later, ideas that arose out of complexity theory and computer modelling, and by this I mean *spontaneous self-organization,* are often used in a way that completely glosses over and avoids any scientific notion of how self-organization is supposedly achieved. Among dynamic systems, patterns may spontaneously emerge out of chaos, and this phenomenon is what has led to the notion of self-organization, as an emergent and spontaneous event.

The self in self-organization is the main problem. There are two meanings of self that can be used in apparent examples of self-organization, and if you can switch back and forth between them, a kind of magic happens due to our language.

One is the self that is the individual in a population that might contribute to self-organization. The other is the self that is the whole organized result of all the supposed organization. So with these two meanings of self, does self-organization refer to organizing being done by the constituent parts or by the, as yet non-existent, resulting whole? Modern science before Levolution has no answer.

Given the computer age, it may not be surprising that networks have found their way into thinking about self-organization. The neural network inspired networks of nodes to become the individual parts and their emergent properties, the algorithmic emergence that makes them form patterns, conjures up the new organization. There is nothing new there.

I submit that without natural selection, which is at its root a thermodynamic phenomenon, there can be no self-organization. Self-organization is a complete misnomer that obscures the point that the phenomenon that these theorists are chasing is entropically functional order.

Self-organization is a vague concept related to the origins of order, of newly ordered systems, or new levels of organization. It has been treated, or at least used, scientifically by W. Ross Ashby, Ilya Prigogine, Ervin Laszlo, Stuart Kauffman, and Eric Chaisson, among many other authors. It now seems to be used ubiquitously in discussions of the subjects at hand, even though it seems to lack a real explanation for the observed phenomenon of actual organization.

The named scientists above are what I would call the thought-leaders of the field. These workers are, in many ways, the pioneers of what I would call the science behind Levolution. Prigogine discovered dissipative structures, and Stuart Kauffman discovered the chemistry of Autocatalytic Reaction Sets that represents an important connection with life, and a whole new level of chemical organization. In that sense, they are allies in the description of Levolution.

However, these scientists are also the main reason why a conception like the misnamed one of "self-organization," has been allowed to survive in mainstream science. These authors' well-deserved stature provides the needed cover and authority, but it is still true that self-organization does not really happen in an *evolutionary sense,* as it actually does in a *developmental sense,* when something like the genetic basis for developmental organization is implied.

It is my obligation to show the error of all these accomplished scientists, but only to the extent that their error is promotion of the idea of self-organization. Self-organization has served as the intellectual placeholder for Levolution in the currently operating paradigm, but it is now time to make the appropriate replacement, and even update the framework.

Self-organization is a concept that grew out of developmental biology, and the neuroscience of brain and mind development, not evolutionary biology. That point is telling because even though "ontogeny follows phylogeny" to some extent, it does so through differing mechanisms and in a very different way.

In nature, a plant's seeds, and even DNA strands, have the ways and means to reproduce themselves programmed within them. Biological development from a fertilized egg is clearly a self-contained operation, and this is undoubtedly at the heart of the confusion. Development is a part of the order-encoding innovation of reproduction of living things, and it is a very different challenge than evolution or Levolution, which are processes of adaptive and entropically functional change.

The challenge in nature that relates to the origin of new levels and new systems, or what we here call Levolution, is to accomplish such events without the information encoding, copying, and programmed activity contained in something like a seed or a DNA strand. Spontaneous ordering, as in dissipative structure formation, is a matter of arranging completely chaotic and un-programmed flows of energy in such a way that continued capture of more energy will result in the continued existence of the form and patterns of its order in its given environment.

There was no seed of single-celled life that contained what is needed to develop multi-celled organisms. The event had to happen through the actions of individual cells in a population or colony. The process is emergent in the sense that it is a population level effect stemming from individual acts, but it is not internally driven. Energy comes from the environment. It is not emergent in any mysterious way either, but natural selection is an emergent phenomenon, and it is an action reflecting the characteristics of the environment onto what may survive. When the complementary parts of something come into individual existence, the whole emerges as well.

Things do not self-organize in the absence of natural selection, but believers in self-organization seem to think otherwise. They tend to believe that all of the important mechanisms necessary to organize a new structure are inherently within or inside it.

Some workers at least recognize that the order comes from the emergent actions of many individuals in a population. This latter point is true and is part of the Levolution story. The situation, however, is that the order that emerges is always wrought by something being subjected to Thermodynamic Natural Selection, and this thermodynamics-driven process, based ultimately on energy pathway selection, is what is creating the order. If one looks close enough, these acts of natural selection will always be found operating in supposed events of self-organization.

Convection Cells Don't Self-Organize

For example, take the classic case of a convection cell in a dish of fluid heated from below, and an assertion that it has spontaneously self-organized.

The individual fluid molecules in such a scenario have differing buoyancies due to their different temperatures, and hence their different densities. Energy is coming in at the bottom and heating the molecules there first. In many fluids, they will typically expand, become less dense, and rise in the fluid column as a result. The relative motion of convection is based on the vectors of kinetic motion resulting from the differential buoyancy of the molecules. Some move up due to their higher heat content. They displace the molecules there, so these cooler ones move out of the way either laterally or downward. Vectors of energy flow are reflected in this differential buoyancy and displacement, and they have effects on neighboring vectors that promote a general separation and alignment of the more widely differing vectors. These become resolved into up, lateral, and down currents. Upward and downward pathways emerge as a "population of vectors" effect, creating the architecture of the convection cell.

The more powerful the differentials, the faster and more orderly will be the convection. Here is the point; random motions or vectors are outcompeted for space by the aligned ones, and they are minimized and naturally selected out in this environment. This is the mechanism of "energy pathway selection" a precursor of Thermodynamic Natural Selection in action. The population of aligned and surviving vectors create the apparent order of the convection cell.

Spontaneous ordering through what has been called "self-organization" becomes spontaneous ordering due to thermodynamic laws. Pathway selection in an environment of increasing energy throughput is key to dissipative structure formation. This example shows that convection cells do not spontaneously self-organize. They are organized by the externally produced energy and its laws of operation. If the vectors were real and discrete entities rather than mere constructs, we could properly call it natural selection acting to make the parts or currents of the roiling cells. It would also be quantitatively approachable.

Self-organization is thus a very scientifically deceptive misnomer, primarily because the effects of natural selection come from environments; and this is exactly counter to the notion of self-organization.

Physicist Per Bak wrote a book called *How Nature Works* (1996) which covers the subject through which he has attempted to advance self-organization. His contribution is the notion of "self-organized criticality" a phenomenon demonstrated by dribbling sand grains and causing natural avalanches of all sizes in piles of sand grains. More easily accomplished and measured, Bak typically employed computer programs to capture this phenomenon as a model. Each run contributed data to the pattern of avalanche sizes found to follow the Power Law in terms of their size distribution. The size variation pattern thus displayed the pattern that reflects the Power Law.

Bak's point, I think, is that with the Power Law dictating that avalanches will come in all sizes, it is really only a matter of time before a critical point, a sufficiently large avalanche, will become a tipping point for a radical departure, a perturbation, a fluctuation, or a bifurcation that somehow mysteriously leads to self-organization. What I could not find in all this is the mechanism of organization that would connect this to the building of a new level of organization.

Fluctuations Again

To me self-organized criticality simply echoes the mystical sounding conceptions started perhaps by Prigogine and perpetuated by others studying dynamics, in which fluctuations, quantum or otherwise, lead to bifurcations and these lead somehow to self-organization. It takes some courage to argue against living Nobel Laureates, but I do not see a theory of organization here at all, whether self-accomplished or not.

Prigogine's own mechanisms for the creation of dissipative structures came from his version of chaos theory, which involves fluctuations. It could be simply my ignorance, but I do not think that fluctuations, which are changes without causes, could scientifically explain anything.

Modern physicists and quantum theorists have used random fluctuations to conjure up absolutely anything, including the large-scale structure of the universe. I cannot follow the herd on that either, and I would put unexplained fluctuations, quantum or otherwise, in the same intellectual scrap heap as self-organization in general.

Many authors have alluded to these mechanisms, often together, and while I have actually read many of them, it is still close to meaningless for me. Spontaneous jumps to different states, while they may actually occur, are a very weak explanation for anything, and simply lack any known causal power for creating order.

The application of dissipative structure concepts to biological systems, however is quite useful, and I was looking for something like them in the eighties, when I found Prigogine's works. The Laws of Functional Order, including discrete dissipative structures, their natural selection, evolution, and Levolution, the processes that rule in all the natural sciences, all stem from them.

Fluctuations, as I said, are basically "changes without causes" that are simply assumed to happen. There is no operational theory about them, and the conception has been allowed to survive unchallenged within physics.

It seems to me that "self-organized criticality" predicts such uncaused changes eventually, as a statistical result of the Power Law, so according to this group of scientists, statistically caused fluctuations as well, have become real, when physical reality has had no say in the matter at all. The two orientations live happily together as a tautology. They may be compatible with quantum theory, but they are powerless to explain the increasing, entropically functional order of the entities we observe; the recursively produced holosystems.

Stuart Kauffman's Mysterious Adjunct

The whole situation of the self-organization is confounded in current science and current thought primarily because even the thought leaders of the field, even those who allow some partial role for natural selection, have maintained that *something in addition to natural selection* is operating in events of supposed self-organization.

Stuart Kauffman, more clearly than most has said this, but none of the leaders of this field give natural selection the profound and exclusive position that it actually occupies in the non-equilibrium thermodynamics of Levolution. Kauffman, at least in the mid-1990's believed in self-organization, and was one of its primary explainers in the realm of chemistry and, in particular, the chemistry of Life. He wrote two books on the subject, *The Origins of Order (1993)*, and *At Home in the Universe (1995)*. His views, as of then at least, are thus explained and available for comment.

Kauffman admits that natural selection is involved, but he does not view the process as the exclusive cause of the growth of order. Something is missing, but Kauffman says he does not know what. However, he does seem to have a model that includes an answer. Kauffman discusses a concept that he calls "Order for Free". He puts this concept in the same vein as the idea that order should be expected.

There are two mechanisms at work in Kauffman's model and these acting together are how he sees the process. One simply springs from nature, to hear him tell it. This is "spontaneous ordering" and the other mechanism is natural selection. Spontaneous ordering, in his conceptions, is at least as important as natural selection in producing the order of the cosmos. This is extremely interesting as it is exactly the way I see it. The differences, however, are in the telling.

Nature's *spontaneous ordering* is the process of its forming dissipative structures. This fact is almost obscured by Kauffman. The other thing is that he views natural selection in the purely biological sense, and not in the thermodynamic sense. If all natural selection required the encoding and reproduction of order, then it would be a merely biological notion, but it isn't.

The main difference is that I must now view natural selection as something much more general than the biological mechanism. For natural selection to act, all that is needed is variation within a population of entities and differential demise. Anyone, or anything, might be responsible for removing some variants from the population, but whatever it was has just performed natural selection and changed the population.

Thermodynamic Natural Selection would restrict this general view to the one where the entities are thermodynamically-defined structures, and the variants are selected for survival based on their energetic properties. This modification is not very restrictive because nearly everything in the universe would find itself covered here, even the biological and cultural entities.

What I believe may have been Kauffman's main difference from the Levolution perspective is very minor. It is fundamentally that he did not see natural selection at work in the formation of dissipative structures and their "spontaneous order".

Prigogine must not have seen it either, but natural selection is what makes the flow vectors of a convection cell become aligned, and it operates by selecting out the randomly-directed vectors. The fittest vectors in a convection cell, or in the swirl at a drain, are those that are not in opposition to the particular alignments that happen to make the energy flow the fastest.

Biology may be viewed as the set of organizing principles that seem to make the right chemical molecules exist at the right place, and at the right time in physical spacetime to make a set of reactions happen.

Biology seems to emerge as a structure to orchestrate the chemical symphony by manipulating physical space with membranes, flows, and other macroscopic, and morphodynamic, means. The processes at work are all 100% chemical and 100% functional, but at about the level of Autocatalytic Reaction Sets, where we start talking about metabolism, life, and biology, we see a higher level.

The mandated thermodynamics of "entropic functionality" and the ideal form of a special "thermodynamic structure" begin to interact again, just as Prigogine noted that they did in all dissipative structures.

The existence of a thermodynamic basis for natural selection stretches back as an idea to the 1920s and quantitative biologist, Alfred Lotka, but the physicists did not pick it up. Energetic ecologist, H.T. Odum also tried to make the point. Maybe the third time is the charm.

While I want to emphasize that natural selection is virtually the sole actor in the phenomena of Levolution, I have come around to the position that indeed there is something in addition to natural selection responsible for the spontaneous ordering observed in so many disciplines.

That thing is not technically an actor, however; it is the structural ideal of the energy flow patterns usually represented in the holosystem structure. Dissipative structures and holosystems are actually an attractor state, like equilibrium. We will discuss this further, but for now let us consider the outlook of a Greek philosopher, and think of it as a Platonic "ideal form".

Plato's Return

The idea here is mainly that this ideal thermodynamic form or structure of a holosystem, which is based on a nested relationship among thermodynamic dissipative structures, is really a set of identifiable functional patterns in the flow of energy that together serve as a target for natural selection of real systems. It is the target toward which Thermodynamic Natural Selection causes hard working energetic entities to evolve.

This phenomenon works because of an ecological reality. Charles Elton was the biologist who first theorized the role-based ecological niche. The role-based niches of any environment are the ways of life, the energy manipulating tricks, the behavioral and functional roles that keep some type of organism alive and able to make a living in some particular environment. Most, if not all, role-based niches are predictably functional in terms of ecology.

A thermodynamically-defined, role-based niche represents an opportunity in some environment, and will predictably be filled by an evolving type of holosystem. At that point, a material occupant has been evolved to fill a role, to be a particular functional component or part of a physical instantiation of an ideal form. Evolve enough of these functional and dissipative parts, and with an eye to the full complement of required component functions of dissipative structures, you will get another whole holosystem. That is the process of Levolution in a nutshell.

We could say here that energy is "participating in the ideal forms" as Plato famously did, but that would be both spooky and anachronistic, in addition to being revolutionary, eye-opening, and insightful. No, seriously, it is not any mythical process of spontaneous ordering; it is the process of natural selection.

Order has to be imposed and enforced upon a variable population of prospective future parts. It takes energy to do that. Structures do not spontaneously attain their ordered state in the normal sense of the word spontaneous. Only in the thermodynamic meaning of "spontaneous," which means energetically downhill, does this make sense. The spontaneity is true in that sense, but the process is not easy, quick, done by itself, or any other meanings of the word.

Order is achieved through the process of surviving natural selection. It is energy's proclivity to conduct natural selection that has shown to us the plan for an ideal dissipative structure, and energy, in essence, targets that structure. Because the most energetic things around to use for parts are the existing systems on the highest level of organization, the target becomes the nested holosystem structure.

Now this kind of talk will be controversial, because it is anathema for biologists to consider any directionality in evolution. That, however, is only because they have yet to be appropriately initiated. Those scientists who adopt Levolution as an operable Paradigm and theory will recognize both the truth and the profundity of this situation.

The beginning of a list of the differing and unique perspectives of the Levolutionary Paradigm might thus include these statements:

(1) *Natural selection targets the holosystem form and is the mechanism behind both adaptive evolution, in the universal sense, and so also the universe's Levolution.*

(2) *Evolution is directional and targeted at increasing the energy flow rates of thermodynamically-defined, ideal, dissipative structures. This is accomplished by Thermodynamic Natural Selection, which causes their continual adaptation.*

(3) *Environments, to which holosystems adapt through evolution, are actually either the inside of a larger holosystem, or represent a region that may eventually become one.*

(4) *The universe is developing in terms of creating ever-larger structures to increase functional order and move energy downward ever faster. The process of Levolution is the process of creating them, one level of functional order at a time.*

Ecosystems are very high order holosystems and the populations of biological species within one are the parts of this holosystem. The species are the parts that have been adapted into such through the operation of biological evolution. Diversity of species is due to the diversity of ways to "make a living" or move energy downward in potential utilizing every nook and cranny in the ecosystem.

The environment essentially controls, through natural selection, what its parts will be. The environment controls what the systems will look like, and virtually all of their details, but fundamentally, the result will be as close to an ideal energy-moving holosystem for that particular niche as possible.

Suffice it to conclude that neither species nor the ecosystems they comprise can self-organize without externally-sourced energy flow and externally sourced Thermodynamic Natural Selection. There is no question here about the direction of their evolution (which is toward closer adaptation and faster energy flow) or their motivation, which is to maximize entropy production and still survive.

Peripheral Concepts of Self-Organization

Self-similarity is very important to Levolution. Autopoesis aside, and fractal mathematics aside, holosystems represent the scale invariant self-similarity of a thermodynamic strategy or pattern. This pattern, the holosystem form, is evident across a range that extends from gravitons to galaxies, and from atoms to ecosystems.

The Levolution Paradigm, and my mission here, is related to all this, but it recognizes and provides an exalted position for the holosystem form, for the process of Thermodynamic Natural Selection that acts on them, and for the creative evolution process and the adaptive changes it makes. Thermodynamic Natural Selection is all it takes to create the parts of a larger whole by differentiating an existing population of dissipative systems.

Previous authors seem to miss how the connection between natural selection and functional order creation is made, but my guess is that they simply have not seen how to draw upon and extend the laws of thermodynamics to the logical framework of energy pathway selection coupled with discrete entities that are dissipative structures to get to Thermodynamic Natural Selection.

I have learned a lot from Rod Swenson, but I do not care for the word "autocatakinetic," used by him, because it's use of the root "auto" or "self" perpetuates the myth of self-organization. Similarly, the word used by Kauffman, "autocatalytic" makes the same erroneous implication. At least in the latter case, the subject is usually restricted to chemistry, where autocatalysis makes sense in the context of a whole, a set of reactions.

Contributing to the general problem of confusion in this area is that instances of natural selection among dissipative structures that are not also holosystems, is very rare. Other than photons, which I will predict here will be found to be electromagnetic dissipative structures, most of the examples of the phenomenon are morphodynamic in character, where gravity, heat, and fluid flows interact. These patterns usually are not durable, although the observance of hurricanes spawning tornadoes is intriguing as an example of the reproduction of a vortex.

Once the thermodynamic ramifications of dissipative structures are applied in biology, one enters a different realm where natural selection is already well-known and accepted. This is the key needed to expand the concept of spontaneous functional order creation. My path was to see natural selection operating in economics, chemistry, among even sub-atomic particles, and then even among gravitational systems.

Natural selection, the differentials in survival reflecting the allocation of energy to the fastest energy-dissipating and degrading pathways, if and only if they occur as variable populations of discrete dissipative structures, is seen for what it truly is; a universal, spontaneous, order-creating mechanism, complete with template.

These phenomena are apparently simply aspects of energy and its laws, which all seem to describe the same project; to maximize entropy in the universe.

CHAPTER 4
THE HOLOSYSTEM STRUCTURE

The way up and the way down
are one and the same.
— Heraclitus

THE HOLOSYSTEM

The holosystem is a thermodynamic structural model of a generalized operational, energy-flowing system that has the properties of being "energy dissipative" like a dissipative structure. Holosystems are dissipative structures; they are recognized, right here, as a variation of this recently discovered thermodynamic version of a Platonic Ideal Form.

The definition of a holosystem includes the definition of dissipative structures, and so half of their uniqueness is because they match the dissipative structure pattern. The other half of their definition, the one that sets them apart from convection cells and swirls at drains, is the requirement that they are composed of other dissipative structures. Holosystems are dissipative structures whose component parts are also dissipative structures.

It is apparent that to understand holosystems one major challenge will be to understand dissipative structures. The nested arrangement of holosystems is the other challenge. On the surface this is simple, but in its depths it is profound. Both dissipative structures and holosystems are thermodynamically relevant structures, and as such, they are unique and special forms. Equilibrium is the only comparable Form, and we know how important that one is.

Dissipative systems display a schematic or general structure of functional patterns of energy flow that are very important to understanding the universe. Holosystems dominate the cosmos, while the other dissipative structures, with the exception of photons, are rarer.

The dissipative structure concept refers to a pattern of energy flows, a model relevant to thermodynamics, which is not much different from a Carnot heat engine. This is the simple steam-based heat engine that was the focus of Carnot, Clausius, and William Thompson (Lord Kelvin) as they first discovered the regularities thermodynamics.

They are energy-consuming, energy-using, entropy-producing systems that are important manifestations of energy in its flowing. They use energy, captured from their environment, to maintain themselves against the usual decaying effects of entropy. What they are maintaining is their form, their order, their dissipative structure, their patterns of energy flow that sustain and define them.

All of the energy consumed must flow through them and return to their environment, but the truth is that the useful part of the energy flow, their work, is their reason for being, their *raison de e*tre*. It is always one use of the energy consumed by a holosystem to maintain their functional order. It is always another use to do external work in the environment. This work, if you are a nested part of a larger whole, is probably work as a part of a larger holosystem.

The dissipative structure's "process of becoming," as their discoverer, Ilya Prigogine (1979), put it, allows or results locally in a faster transfer of energy.

All of the energy that holosystems or dissipative structures use in maintaining their structure, their order, or in doing external work is eventually degraded to another form, or simply dissipated to the environment. This is also generally true of the energy captured but not used by the holosystem at all. It all contributes to entropy production.

Holosystems increase entropy production. The energy flow rate, always downward in potential, increases in proportion to the increase in energy consumed as holosystems are created on successively larger scales.

Two things always emerge along with each new level or holosystem type, and these are (a) the speed of the entropic energy flow increases, and (b) the functional order of the universe increases. The universal order is functional order, and it is the result of the dissipative structure form being repeatedly implemented in successive levels of holosystems to ever-higher order and scale in the universe.

Dissipative structures are common. All these are discrete entities that make energy move faster through themselves than it would have if they had not formed. That increase in energy flow rate, which translates directly to entropy production, is the ultimate cause of their origin, and it is the cause of their continual adaptation. Energy acts as if it wants to flow faster. It always does so whenever it gets a chance.

Holosystems are all examples of such thermodynamically similar dissipative structures, but they also have further qualifications. Holosystems are always composed of populations of smaller dissipative structures, or smaller holosystems, by definition. This makes their energy flow pattern distinctive, and it unifies them all into a very similar thermodynamic model. This model is found at each level in the nested structure of the universe's levels of organization, and it traces the path of Levolution.

A holosystem is a kind of generalized pattern of energy flow that is obedient to the laws of classical thermodynamics, and each type of holosystem has a particular role, function, or capability with regard to some form of energy. This functional specialty, this manipulation of some energy form for particular uses, is a system's *functional order*. This particular property, which is a generalization of the details, I will only tend to discuss in generality, because thermodynamics is a general set of laws. Functional order always makes whatever form of energy the holosystem can capture and manipulate flow and dissipate faster than before that type of holosystem existed.

An important thing to realize is that almost everything is a holosystem. In this book I will explain how they arise, how they disintegrate, and how everything from sub-atomic particles, to gravitational systems, from atoms to living things, cultures, and ecosystems are all holosystems. Holosystems share basic thermodynamic phenomena, and because they accelerate energy, they have been allocated, by their environment, an earned fraction of the energy of the cosmos. Energy in its flowing has built up one level at a time to produce the material aspects of the universe we know.

Previous authors, including Prigogine, have noted that living things are dissipative structures, and this is profoundly true. Few authors, however, have tried to make the point that gravitational structures, or the particles of high-energy physics are dissipative structures. I will support the fact that virtually all of the obviously ordered structures of the universe are holosystems, and this points to both the likely truth of the conjecture, and another principle of importance.

The "order" of dissipative structures is the only thermodynamically legal way to spontaneously produce or increase functional, energy-moving, order in the universe. If you observe this kind of order, it should be immediately suspected that you are looking at a dissipative structure. Further, holosystems in their distinctive nested arrangement represent the typical way, the common way, that energy finds its way to lower levels of potential.

In the cosmic view, simple radiative dissipation of energy into space, and simple energy conduction through materials did their work long ago in the first moments. We live in the era of holosystem building, and the "building" process, which is the Levolution of holosystems", is the central energy-moving phenomenon of the universe today.

A holosystem may seem material, just as you and I seem material, but like all holosystems, we are really more accurately identified with the flowing energy itself. A holosystem is really a pattern in energy flow that speeds up energy flow. Take away the energy flow and we are dead. The pattern holds for us and for all holosystems. The holosystem may be the thermodynamic model of life itself.

We, like all holosystems, are composed of smaller holosystems; our cells. A long time ago, primitive cells called prokaryotic cells merged into modern cells and became parts called organelles inside them (Margulis, 1970). Before that, autocatalytic chemical reaction sets had merged into those primitive cells (Kauffman, 1995). Before that could happen, a plethora of chemical species combined through the action of sharing and transferring electrons. Stable atoms had to first form from nuclei by combining with those electrons. Before this, nuclei had to be forced together by outside gravitational forces to make the various elemental nuclei inside stars. Before this, nucleons, like protons and neutrons, had to form, each from precisely three quarks.

Nature has left a clear trail of its Levolution, and it is the trail of the levels of organization observed among holosystems. This is the observation that Levolution explains. Levolution is not, therefore, unproven or speculative. It is proven primarily by the existence of those levels.

The Construct

The holosystem began in my mind as a theoretical construct, a contrived schematic pattern, based on a set of generalized system properties contributed by others; properties that are universal. They are universal because they relate to energy flow, and the thermodynamics of nested dissipative structures. Thermodynamic laws pertain to all forms of energy, not just heat as some people will try to tell you.

Figure 3 is a schematic diagram of energy flow through a holosystem. It has been greatly simplified to show only one representative part of the nested whole. This is a necessary simplification, because some holosystems have millions of parts.

Note that there are two black circles, each of which represents a holosystem on a particular level of organization, two of which are shown. The black circles are also the holosystem boundaries. Note how the smaller one is an internal part of the larger one. It is also a whole composed of parts itself, but they are not shown. The key is to recognize each level as both a whole and a part; a whole-part duality.

Note also that the Figure contains (1) the general system requirement of a boundary condition, (2) the dissipative structure pattern of energy flow, involving capture, use, and dissipation, and (3) the part-whole relationships of a nested structure.

Note also that these three universal functions of dissipative structures; (1) energy capture, (2) energy use, and (3) energy dissipation are also being performed by the parts. This results in the need for three more universal functions for holosystems. These are the part management functions of (4) distributing energy to them, (5) of managing their work output, which will become the work of the whole, and (6) of dissipating the entropy that the parts produce.

These six universal functions are always found among holosystems. In general, this type of pattern is known as *scale invariance*. It is a mathematical property of most fractals.

The energy flow pathways or functions connect the levels of organization with flows that go both upward and downward among the levels of organization. What we have here is a schematic of the generalized energy flow plan for virtually everything in the universe.

The energy captured by a holosystem can flow and be used in many ways, but it is always logically and conceptually separable into several downstream components. The first conceptual fork in the path is one that represents the manifold distribution or allocation of energy resources to its parts. These may be few or many. The energy resources distributed from one level will be distributed down to parts, which they will view as energy capture, but it will come back upward to the whole as a contributing manifold of work done by parts performing their roles or functions for the whole, and as the entropy that they dissipate. Notably the entropy of parts must be handled by the whole. The whole is their environment.

The second type of fork in the path of energy is the entropy-law-imposed separation into the two portions, useable and non-usable energy. Usability is a value judgment and is relative to the system involved, and varies among them. For our purposes in qualitative description, we simply need to know that there will always be these flows operating. They must connect and quantitatively balance, satisfying the law of the conservation of energy.

Usable energy is the portion that the holosystem can put to use in creating and maintaining its order, or doing work in its environment, but it also includes energy that one could perhaps say is emergent. Many of your muscle cells must contract in unison for you to even lift a pen. Keep in mind that these forks in the path of energy are conceptual ones. They will be there, but they are not necessarily easy to separate and find. The energy actually used by a holosystem, or by its parts, will be dissipated as it is used, so all of the energy that a holosystem captures will ultimately be dissipated. However it is true that some of it may be used for the purposes of the holosystem first, and some of this will be useful to the higher level, of which it is a part.

Among holosystems, and all dissipative structures, creating internal order is the means by which they speed up energy flow. Order creation may be considered as their very purpose. Entropy is like the price paid for the order of a dissipative structure, which is the means the dissipative structure has used to increase energy flow.

All energy must be accounted for, and to maintain itself as a dissipative system, the entropy dissipation must be sufficient to pay for the order. Some people call this the "negative entropy" or "negentropy" but I call it "functional order". It is represented within all dissipative structures. That apparent cosmic deal has been made between energy and order in every dissipative structure and holosystem here.

Holosystems speed up energy flow wherever they have been created (by Levolution), and are maintained in a state that is well adapted to their environment. Adaptation to the environment means that they can transfer more energy to lower potentials, faster. This logic is characteristic of all situations involving holosystems.

Holosystems undergoing Levolution are essentially adapting to larger and larger scales or dimensions of energy flow, allowing huge amounts of power to flow through huge holosystems. Galaxies are almost certainly holosystems. What started out as pure energy in the Big Bang has now become gigantic gravitational holosystems of matter.

Six Functions

In particular, holosystems are energy flow patterns representing the six universal functions enumerated above. These are six energy flow segments which must be there for the system to operate.

The model is qualitative and entirely general. Only the particular details of these six flows, the particular ways of organizing to make a living, are left out of the holosystem Structure schematic, but the six general functions of such systems always exist. For many particles this is currently an unproven assumption, but it is a reasonable one because, for every other holosystem, it can be demonstrated. We understand intuitively that it the six functions apply to us and our organismal physiology, but the pattern is universal.

The universal holosystem functions may be said to be due to the ubiquity and shared characteristics of the "struggle for existence" as a nested dissipative structure, or henceforth, as a holosystem. The functions relate to challenges that one might say are ecological in nature, because they relate to well-known system-environment relationships.

The Levolution Paradigm itself makes a tweak to the characterization of such relationships by viewing "environments" as the "whole" side of part-whole relationships. Fitness for holosystems thus becomes fitness as a functional part of a larger whole.

Environmental niches for holosystems become part-roles that they may play in order to obtain energy and resources from the whole. The Paradigm does not exempt biology, so even Darwin's evolutionary theory must adapt to accommodate the perspective that the ecological community is another systemic whole. The ecosystem's contained parts are evolving species of organisms.

As I continued trying to apply the theoretical construct I had made, I found something more profound. The model gradually, as I sought to apply it across the board, became more palpable as a good model for any nested dissipative structure, but beyond that, I could postulate generally that this model was essentially *the* model toward which the process of Levolution always worked in creating new holosystems. Evolution similarly works to keep these entities tuned as well-adapted holosystems, to fit in their part roles of a "whole" or their "environment."

In other words, the holosystem began as a creative construct, but is now posited as a "real" and generally thermodynamic model, and I think it is destined to become part of thermodynamic law. As the next chapter will reveal, the First Law of Order, as I enumerate them, describes dissipative structures, like Be*nard cells and swirls at drains. A proposed Law of Functional Order describes the holosystem structure or model, as dissipative structures composed of lower-order ones that have aggregated and now live together. A holosystem is fundamentally a whole system made from a population of smaller, lower-order, dissipative structures or holosystems. The aggregation process itself is Levolution.

Energy is the common denominator of the universe. Its laws are universal, and that is why there is sense to be made of the universe. The so-called "forms" of energy are nothing but the transformations made in the primordial energy that arrived as a Singularity before the Big Bang. The transformations are primarily made today by holosystems that dissipate and degrade the energy. Dissipated and degraded energy seems to have a different form, exhibiting less concentrated potential for sure, but the energy is not destroyed, it is simply changed and spread out. Another holosystem may still use it, and that is a key point to remember.

The holosystem structure regards the processes that energetic systems in the universe have in common. Other workers discovered these properties for the most part, and you may be familiar with them. The holosystem model is a combination of three important "general system" theories that together create a general description that fits the naturally occurring, nested, whole-part dualities in the universe.

The synthesis that creates the holosystem Model combines the ideas of Ilya Prigogine, Ludwig von Bertalanffy, and Arthur Koestler, each of whom described systems very uniquely, but nevertheless did so in terms of their most general and universal principles. Prigogine gave us *Dissipative Structure Theory*, which as noted above is the only known spontaneous source of functional, energy-transferring order in the universe. Bertalanffy gave us *General System Theory*, which gives holosystems interrelated parts, and the needed system-environment, and boundary condition perspectives, and Koestler gave us "whole-part dualities" and a name for the entire nested structure in his scope, which he called the *Holarchy of Living Nature*. I combine these two of Koestler's notions with those of Prigogine and Bertalanffy, and add my own ideas about thermodynamic fitness and Levolution, as the cause of the levels of organization in the grand holarchy of the holosystems.

Levolution is the process that creates new holosystems as dissipative structures formed out of those on the level below. The generalized construct of holosystems thus contains a thermodynamic qualification from Prigogine, a system qualification from Bertalanffy, and a structural qualification from Koestler. Holosystems create a new story about what the universe is actually doing. I think that physics is waiting to hear this story.

Even while entropy inexorably moves energy downward in potential, and disrupts what I call thermal-kinetic order, the Levolutionary production of new holosystems goes the other way; it builds functional order. Dissipative structures in flows of energy make the energy flow faster. Levolution, in creating new types of holosystems out of holosystems on the level below, has built all the systems of the cosmos, from particles to galaxy clusters for that entropic purpose. Energy's mandated entropic tendency to create dissipative structures has powered it. It is obvious on some levels, and while it is less obvious on others, the overall pattern clearly suggests the universality of this phenomenon on all levels of organization. I think you will be as surprised as I was when you become convinced that this is the case.

Holosystems are the *exclusive* subjects of the new theories presented here. There can thus be no question as to what these new theories pertain. They pertain to all holosystems and only to holosystems.

Holosystems conceptually, structurally, and schematically, represent a *common pattern* in energy flow that is observable in every naturally occurring durable dissipative system that has come about through Levolution. The commonality of the pattern among so many systems is itself very important. It is this commonality that makes the construct work, but interestingly and more importantly, it is also this very same commonality that exists between a "holosystem as a part" and a "holosystem as a whole" that allows Levolution itself to work.

The "holosystem pattern" of energy flow is common to all such systems because energy flowing always drives it toward the dissipative structure pattern to speed its own flow. That is the most important fact to focus upon! Levolution, as we will see, operates through Thermodynamic Natural Selection acting to sculpt out a new type of holosystem. The holosystem form is a "state attractor" but it is not Ashby's proposed equilibrium state attractor. Levolution is a property of energy flowing far from equilibrium. It is a non-equilibrium thermodynamic principle that has resulted in our cosmic structure.

The word *holosystem* is from the Greek word for whole, "holo" and "system" which, since Bertalanffy, is an operating, functioning, entity with parts, and with a discrete separateness from its environment. A holosystem is a whole; it is composed of parts, and it is usually a part of a larger whole.

A Universe of Holosystems

The universe is full of differing types of holosystems. They were listed in Chapter 1. This book is about how they originated, and how they evolve into continually adapted part roles. I have concluded that there are exceptions. Not everything is a holosystem, but those that are not are generally nondescript populations of holosystems. Everything down to and including gravitons, the prospective sub-atomic particles, and everything up to galaxies and the large spherical voids in the large-scale structure of the cosmos, among gravitational systems, are holosystems.

Experimental proof of this may not come for a while, especially for particles or gravitational systems, but for chemical, biological, ecological, economic, and cultural phenomena, and really for everything subject to these thermodynamic laws and principles, it is the logic chasing the observations, rather than the other way around.

It is really the "logical momentum" if there is such a thing, of the finding that a universal form of evolution has swept, and is sweeping, across the conceptual landscape of holosystem types that has carried the theory of Levolution down into the realm of sub-atomic particles. Levolution is more or less obvious and accepted in biology, and the holosystem pattern is plainly visible on these levels.

At the chemical level, self-catalyzing families of reactions operating in the soup of primordial earth probably evolved into primitive biological cells. Stuart Kauffman, who in my mind has discovered the missing Levolutionary link that gets us to the living levels, has studied this transition from chemistry to biology productively. Life's definition is problematic however, and energy flowing in holosystems is very close to the thermodynamic definition.

The sub-atomic levels of organization are more difficult to see as holosystems, but we can clearly see the hallmark of a nested holarchy of structure, as is created by Levolution. We are close here to solving the riddles that will allow us to see particles for what they really are. If we are prepared for it, we may see it.

I know that the Theory of Levolution is, to some degree, unproven and incomplete. While it does provide a new explanation for gravity and the origin of earliest particles, that part is almost pure logical speculation. It is perhaps less speculative to note that three quarks came together and Levolved into the proton. I think physicists would agree on that. Protons and neutrons did come together to make atomic nuclei. That should not be controversial. Wherever Levolution is incomplete, it is because particle theory is incomplete and the story of energy is incomplete. This is not necessarily a failing of Levolution. Levolution, to the contrary, provides particle physicists with things to look for. I'd bet there is a gluon in the middle of a graviton somewhere, but more on that later.

Encouraging signs are evident. Particles are observed to absorb and emit energy. The Standard Model seems to indicate several levels of organization in the sequence: quark, nucleon, nucleus, and we know the next level is the atom. Most of the holosystem criteria are already in place. Consideration of the rest of them blew the lid off my simple little book as my own eyes were opened by "following the energy".

Entropy and Dissipation

The dissipative structure is an important construct in thermodynamics. It is absolutely central and crucial to understanding our subject. We will delve deeper later, but this is the construct that allowed biologists to understand that the living order visible in biology was not thermodynamically illegal, and not a case of the entropy law being temporarily "retarded" or "reversed." Dissipative structures create order, so long as they pay for it with captured environmental energy, and display relatively faster entropy dissipation in the overall picture.

Energetically dissipative holosystems have been created, and are continually adapting, as discrete "energy accelerators" that speed up, and maintain the speed of a flow of energy in the universe. The hardest part to grasp here is not the universal applicability of the holosystem model. I can convince you of that. The strangest and most profound part comes from the fact that all functional order creation in the universe is the result of the headlong rush of energy toward entropic doom. We have been mistaken about entropy for many years. It is not mainly about the destruction of order, which it does in simple situations; it is about the creation of order, which it does in ubiquitous dissipative holosystems.

Entropy is, in effect, the driver of order creation. By making energy flow, inexorably and mindlessly, downward in potential, the Second Law of Thermodynamics creates an opening for order, if and only if, that order is functionally aligned with entropy and can also speed up energy dissipation. Structural order, which is to say arranged, channelized, non-colliding, energy flow pathways, increase the energy flow rate in holosystems and that is the only reason why functional order exists at all. Energy flowing creates functional, holosystemic order, and this entropically functional order increases the rate of energy flow.

It becomes abundantly clear that the purpose of entropy, an inviolable natural law, is not really to destroy all the order in the universe, per se. It is simply to dissipate energy and reduce all energy potentials. Outside of holosystems, this manifests everywhere as decreasing order, but within holosystems themselves, it manifests as increasing functional order. It is as if entropy, and so the whole universe, destroys all other forms of order, but allows, and even creates functional order.

SYNTHESIZING THE HOLOSYSTEM

Whether we know it or not, nearly all scientists study **holosystems**. Holosystems are a generalization of the common properties of real, naturally occurring systems. Simply put, they are a synthesis I created from:

- Dissipative Structure Theory
- General System Theory
- Levolutionary Holarchy Theory

Let us now look at these in a little more detail. They each contribute essential concepts to the holosystem model.

Dissipative Structure Theory (DST)

Holosystems are, thermodynamically and fundamentally, dissipative structures, but they are a subset, and a further level of classification of them that I have based solely on their nested character. The distinctions are very important, mainly because they distinguish between dissipative patterns of fluids like convection cells and the dissipative metabolic and chemical patterns of biological cells. Being a whole composed of parts becomes more important to recognize when we survey the scales upward or downward in size, but being dissipative in thermodynamic character is a more general characteristic.

Dissipative structures can be relied upon to do three things. They capture energy from their environment, they use energy to do work and create and maintain their internal order, and they also produce entropy or dissipate energy faster than before they existed. They speed up the overall energy flow through themselves.

Their internal "order" is the set of energy flow patterns that they must create and maintain in the face of entropy, and in the face of whatever challenges their specific environment throws at them. The order is supportive of their dynamically stable existence. They exist far from equilibrium.

The well-known state of thermodynamic equilibrium is a basin of dynamic attraction, toward which many things tend when energy is unavailable. It is characterized by a lack of energy flowing, and the obliteration of gradients in energy potential.

A Dissipative Attractor State

My own interpretation of dissipative structures, which is derived from the addition of ecological theory, is that the dissipative structures of nature, still spontaneously, tend toward another type of state attractor. It is a local high point in the rate of energy flow. The dissipative structure is a state attractor, as is equilibrium, but it is a state of maximal energy flow and entropy production, driven by an overabundance of energy. As shown in Figure 4, a dissipative structure state is a "dome" of attraction in terms of dynamics, to the point of maximum energy flow representing a flow that is augmented by entropically functional order. We will discuss the Law of Maximal Entropy Production shortly, but suffice it here to say that the structure is in line with that motivation. While energy flows at a high rate in this attractive steady-state, the high flow rate is also maintained over time, and this state is not simply defined by the temperature of the environment.

Figure 4—Ideal Thermodynamic Structures

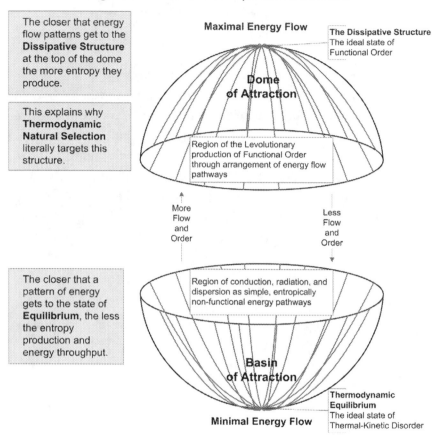

The closer that energy flow patterns get to the **Dissipative Structure** at the top of the dome the more entropy they produce.

This explains why **Thermodynamic Natural Selection** literally targets this structure.

Maximal Energy Flow

The Dissipative Structure
The ideal state of Functional Order

Dome of Attraction

Region of the Levolutionary production of Functional Order through arrangement of energy flow pathways

More Flow and Order

Less Flow and Order

The closer that a pattern of energy gets to the state of **Equilibrium**, the less the entropy production and energy throughput.

Region of conduction, radiation, and dispersion as simple, entropically non-functional energy pathways

Basin of Attraction

Thermodynamic Equilibrium
The ideal state of Thermal-Kinetic Disorder

Minimal Energy Flow

The dissipative structure is a very special energy state created by the specific environment that such a structure or system finds itself within. It is a dome of attraction, rather than a basin of attraction, but it similarly represents an ideal thermodynamic structure. Being reversed from the equilibrium thermodynamic perspective, one might wonder what drives the system upward in terms of energy flow rate. The answer is the pressure exerted by the Maximum Entropy Production Law, and if, and only if, we are talking about a population discrete and variable structures, the Law of Thermodynamic Natural Selection.

The subject high point in energy throughput and entropy production is, importantly, also a peak in terms of "fitness" and natural selection.

Achieving the steady and ideal state of the dissipative structure in a given environment is a very pro-survival achievement. Technically, it is true to say that a dissipative structure is an ordered state that matches up with a set of environmental parameters in a multi-dimensional niche space.

The multi-dimensional niche space was the conception of famed ecologist, G. Evelyn Hutchinson, and the concept of a landscape cre ated on this conceptual space was the conception of geneticist, Sewell Wright. The conceptual region over which the "dimensions," or environmental parameters vary, are what ecologists call a "fitness landscape". It is a set of environmental circumstances among which some type of evolving system might be driven by natural selection. The fitness landscape defines the realm of the niche and the form in which an occupant might continue to exist. With the marriage of thermodynamics and natural selection, which is essentially our central subject in Levolution, we have arrived at this important notion:

The high-energy flowing, steady state of the natural dissipative structure is a state attractor similar to Equilibrium. The higher rate of energy flow in this state is attractive because a higher energy flow rate, consistent with survival, is always an advantageous characteristic in terms of Thermodynamic Natural Selection.

So, the energy flow rate of a system is driven upward through "exploration" and natural selection operating to drive it toward the highest points in the fitness landscape, as current biological theory holds, but these coincide with energy maxima and with the achievement and maintenance of the dissipative structure pattern in the flows.

Dissipative structures persist as long as they can continue to do the three noted functions above, and as soon as they don't perform these universal dissipative functions, as I call them, they cease to exist as a stable dissipative structure. They do not last forever, and their "mortality" can even be an important feature. The mechanism of their mortality is also important. It is ultimately always a fundamental failure to maintain thermodynamic stability as a dissipative structure, a point to which we will return.

Prigogine's concept of dissipative structures solved a long-standing conflict between thermodynamics and biological evolution. The conflict had originated with William Thompson (Lord Kelvin) as soon as Darwin published his work, only seven years after Thompson had coined the word "thermo-dynamics". Order was either being eroded by entropy, or created by evolution. Which was it? Dissipative structures finally allowed the needed integration between biology, where order obviously grows, and physics, where entropy often works to obliterate order.

Dissipative structures, distilled down to their thermodynamic essentials, exhibit an important and widely observed pattern of energy flow, reflecting as it does, the capture, use, and dissipation of energy through an obviously ordered system. The dissipative structure performs work, and it works to locally, and temporarily, thwart the effects of entropy, by maintaining the order of its own structure. Everything from tornados, to living organisms, to convection cells in fluids, are dissipative structures, but they are not all holosystems.

The scientific conception of biological organisms as "dissipative" systems actually pre-dated Ilya Prigogine's description of the thermodynamic concept of dissipative structures. Back in the 1920's, quantitative biologist Alfred Lotka (1922) knew that entropy was pointing in a different direction than living systems. Ecologists who later studied energetics, like H. T. Odum in the 1960's, reiterated the notion that natural selection is of a thermodynamic nature.

When I took basic biology courses in the 1970's, the divergence between life's order and entropy's chaos was considered an "exception" that living things somehow enjoy. The biologist's subject life forms were clearly very far away from any notions of chaos, and far from any equilibrium too. We biologists now appreciate the substantiation provided by Prigogine's conceptions.

The understanding and acceptance of dissipative structure theory has truly broken the chains holding back ecology and evolution, and turned them loose to improve our understanding of the rest of the universe.

Physics, via thermodynamics, is invading biology, and biology is invading physics right back. The two disciplines are coming together rapidly here. Particle physics, gravitational physics, and physical cosmology will all get a big dose of ecology before we are through.

Dissipative structure theory did not explain biological order in any detail. It only "allows" order, by showing how the entropy law would be obeyed, even as order was being created and maintained within the system. Prigogine, and many authors since, have noted that order creation by dissipative structures is "spontaneous." They are using the thermodynamic definition of that word, which means thermodynamically "downhill."

In other words, order creation is not explained at all by the theory of dissipative structures, the phenomenon is simply allowed by it. Levolution is what is needed to *explain* the spontaneous process that actually creates the order.

General System Theory (GST)

Holosystems are partially defined via Ludwig von Bertalanffy's General System Theory, or GST (Bertalanffy, 1963). Holosystems are "general systems," which means they are discrete and distinct from their environment, and have boundary conditions that separate them, but still let energy flow in and out. They are open systems that have inter-related parts, which is a fact we will emphasize later. This is a simple view of GST, and it is all we need here. Feedback and control are not universal and may not be a feature of all systems.

This generalization dovetails nicely with dissipative structure theory. We have gained a slightly improved picture, with properly bounded systems, related parts, and a distinct environment to ponder in an ecological context. The young science of ecology adopted GST wholeheartedly, and the science could even be defined as the study of system-environment relationships. This is close to what we need, but a modification is needed.

Bertalanffy set out on an interesting project to describe the general system. It is now much clearer to me why he undertook the project.

I believe he was striving to get to the construct of the holosystem. Historians can figure it out, but I believe he was close. Once the general characteristics of the subject systems are known, their phenomena come into view more clearly.

Levolutionary Holarchy Theory (LHT)

Holosystems, levolving from parts as they do, are usually composed of other holosystems, but the very first one was composed of dissipative structures. It is a definition of the model that all holosystems are dissipative structures composed of parts that are dissipative structures. That is not true of all dissipative structures. So, while all holosystems are dissipative structures, not all dissipative structures are holosystems.

Holosystems are an abstracted generalization of the very large set of real, nested, dissipative systems in our universe. You are one. I am one. Each of our cells is a family of interlocking, self-catalyzing, chemical reactions within a boundary, and each of these cells is a holosystem. Our culture, and the planetary ecosystem that contains our culture, are both holosystems. The gravitational systems in the cosmos are holosystems, and I will soon explain that. Even the particles of the Standard Model of Particle Physics, the quarks, nucleons, nuclei, and atoms are holosystems.

The universe has distributed energy potential all over the place, but it operates under a thermodynamic law that mandates that energy potentials should all be reduced. The universe is on that mission, and it accomplishes it, in large measure, by creating holosystems. Some energy simply conducts or radiates into dissipative oblivion and obscurity, but much of it now flows through holosystems.

Holosystems create all the magic of the universe. They represent structural and functional order on the increase, but they are doing exactly what thermodynamic law dictates they do. They mop up pools of energy potential and reduce them, dissipate them. They are the energy-burning systems like you and me, working to create order in the universe, and they all work very hard doing just two things, dissipating energy and creating entropically functional order.

Energy is flowing in the universe, and has since the Big Bang. It flows as it does, thanks primarily to holosystems. Holosystems all perform the dissipative structure functions, the energy-related functions of energy capture, use, and dissipation. Holosystems also perform several more energy related functions related to the management of the energy both for and from their parts. Moving energy downhill is truly what the holosystems, and even the other dissipative structures, are doing in the universe. It is their reason for being. They increase the overall rate of energy flow in the universe. They organize the flow patterns of energy, create channels, and reduce collisions.

At the organismal level, some holosystems are observed to gain the psychological property of *intention*. Their behavior, our behavior, is guided by internal ideas, but these ideas reflect their reality as a holosystem. They explore their environment, use their intelligence to find the hidden pools of energy potential, and they create, through the construction and maintenance of their energy flow pathways, the property of entropically functional order, holosystemic order, in their cultures.

We are free to think, but we had better be thinking about our survival or we are off the track. Both Universal Evolution and Levolution act upon holosystems almost exclusively. This discipline has forced the holosystem definition to be precise.

From quarks to clusters, we can detect them with current technologies. As most people have noticed, the holosystems of the universe are arranged like the nested Russian dolls that fit inside each other. We tend to make scientific disciplines out of the various levels, and rarely compare notes between them. That is why we have reached this year and are only now realizing what the universe is all about. The sociological separation between the siloes of the disciplines can now melt away. The rift between physics and biology can heal. Levolution is a true story of unification, of what the universe really is, what it is doing, how it got this way, and where it is going.

There are only a few principles involved in the Levolution Paradigm, but they apply very broadly. There are two related theories here. They are both "grand" theories, and together they do create a theory of pretty much everything, but they are simple and non-mathematical. That is perhaps their biggest problem.

Even if you are not the mathematical type, you will still find these principles understandable, logical, even intuitive. They have mostly come from ecology because the system-environment model is so very relevant for dissipative systems in the energetic universe, but even ecology will be challenged to change. Ecosystems are holosystems, and Darwin's evolution is but an instance of a more universal natural law related to energy and order.

Not shown in the holosystem diagram (Figure 3 above) are some of the lesser-known principles of the energy that is flowing. The lines representing energy flow are constrained by order, but they are in a big hurry. Holosystems, as dissipative structures, only formed in the first place because the energy that was flowing at a lower level of organization "found a way" to make it flow faster, and that caused the pattern of the higher level to develop. The actual theory of Levolution as a process goes into how this occurs.

Maximum Entropy Production

Energy may be mindless, but among all of the natural laws that I have encountered, the entropy law of energy stands alone in the fact that it apparently gives energy a purpose. Even Francis Bacon, whose guidelines for empirical science nixed the notion of using teleological arguments in science, would probably agree that energy has the not-so-profound purpose of flowing downward in potential as fast as possible. It is not a "mental" purpose. It is a natural law-given purpose, and to my mind, that makes it perfectly all right within science.

The direction and the speed of the downward flow of energy represent two different principles. Entropy itself, the Second Law of thermodynamics, provides the direction and points from higher to lower energy densities, levels, or states. This is also the direction of time. The additional mandate to do this "as fast as possible, within constraints" is another principle, the Maximum Entropy Production Principle, which is best understood through an analogy that I have borrowed from Rod Swenson, (1978).

It warrants repetition. Think of energy flowing to lower potentials as if it were rainwater collecting in little streams and flowing down a mountainside. The details of the topography, the irregular contours of elevation, are the detailed constraints that will determine the pathways of rainwater flow. The water, due to gravity, will simply seek its lowest level and proceed downward. According to this law, it will also automatically, and without thinking, take the steepest, deepest, and widest available paths; the ones that allow the water to move downward at the fastest rate.

Water on a mountainside will not wait at all, will not optimize, but will flow immediately downward, always following the "selected" pathways that present themselves as the steepest and widest available, the path that will get the water down the fastest. Unless, that is, it gets trapped in ponded situation, bounded by constraints. Energy is the same as the water in all respects. It is probably mindless, but it has an apparent purpose anyway. It uses what is available, and it operates within constraints.

The Maximum Power Principle

The Maximum Power Principle is one that began in ecology with Alfred Lotka, and became more refined by the ecosystem energetics work of H. T. Odum. I am noting this biological concept here because it is clearly related to the Maximum Entropy Production Principle, and we need to explore it.

These naturalists observed that living systems all behave in a way that seems to fill up ecosystems, grow grass in the cracks of the sidewalk, and leave no wasted resource. Nature, as we know, is prolific, tenacious, expansive, opportunistic, and hard to stop. It expands into the deepest crevices, and out to the farthest edges, and into the most extreme environments. I am not trying to romanticize nature, or resurrect eighteenth century vitalism. This is an accurate description of the way living systems act. Nature's living systems all work hard, and the rate at which they do work is always maximized, within the environmental constraints. Competition for resources, and natural selection based on its outcome has resulted in this situation.

Because work per unit time is the definition of power, this principle is called the Maximum Power Principle. Everything out there in nature, even though it may not always seem that way, is working as fast as it can. There may be nothing special about biology here.

We must investigate whether the principle might operate for all holosystems. Gravitational holosystems are probably making all the gravitational attraction force that they can muster. When water flows downhill, it does not wait for optimal conditions. It immediately flows as fast as it can.

The Maximum Entropy Production Principle and the Maximum Power Principle are related. However, neither H.T. Odum's statements nor Lotka's actually present a well-formed, completely correct statement of the principle here that belongs with other Thermodynamic Laws. Rod Swenson's Maximum Entropy Production Principle as modified here, does.

In using the concept of maximum power rather than maximum entropy production, the Maximum Power Principle misses the distinction between *power* and *efficiency* in a thermodynamic system. Often in nature, either power or efficiency is maximized, depending upon which is more fit in a particular environment, but what is always maximized, regardless of that distinction, is entropy production over some amount of time.

Entropy production means the sum total of all the energy throughput of a thermodynamic system. It naturally includes all the energy passing through the system that does not produce work, but it is also true that all of the useful energy that is put to work also degrades to increased entropy in the environment. In other words, the principle of maximum entropy production can be understood as *maximum energy throughput* within constraints. That brings such systems into line as dissipative structures.

Holosystems and Functional Order

As I said, this book captures a paradigm shift. It is a mind-changing set of ideas that pictures the universe as a single, nested structure of similarly ordered entities. They are all involved in the thermodynamic project, and apparent mandate, to transfer energy to lower potentials as fast as possible within constraints. Each one does so until it becomes a part of a larger system, or not, and generally performs a universal set of energy related functions until it stops working.

This will go on until the universe reaches equilibrium or entropic doom, a state where nothing else can happen. Holosystems speed up the approach to entropic doom. These entities produced by events of Levolution are produced to increase the dissipative, degradive, downward flow. They represent a thermodynamically ideal structural pattern that is found among sub-atomic particles, among atoms, molecules, reaction sets, cells, organisms, cultures and ecosystems. And as we will see, it is also found among every gravitational body in our universe.

The Levolution Paradigm is about the primary drivers of Universal Evolution and Levolution, which is the entropy law and the additional law that mandates the maximization of entropy production. Even more importantly, however, the Paradigm covers the secondary and resulting trend, which is the functional order that all the dissipative structures, but mostly the holosystems, in the universe are creating. It is a temporary order. It lasts in humans only for a single lifetime, unless it is communicated and passed along. There are no holosystems that last forever.

On the Survival of Holosystems

There is an important abstract concept that should be noted about holosystems. Energy and the regularities or laws of nature have created holosystems for the entropic purpose, and has done so within a given environment, within a particular set of constraints, but that purpose is not achieved when they are not in existence. *The functions of holosystems are not performed if they are not there to perform it.*

In other words, the very same entropic purpose that gives holosystems their reason for being is what makes holosystems tend to remain in place. At the highest of organizational levels, where we can think about holosystems with the instinct for survival, we see no problem with the notion that holosystems seek to remain in their thermodynamic niche and do their work. Among particles, molecules, and gravitational holosystems this becomes something of a problem because they have no capacity to think such things.

The behavior of all holosystems seems to be thermodynamically governed by a default tendency to remain in place and survive, if the environmental conditions allow, and all without a single thought. Where a biologist might speak of survival within a niche, a physicist or chemist will speak of maintaining stability within environmental constraints. When any holosystem exists, it inherits a sub-purpose from the primary purpose of entropic drive, and that is to simply continue, within constraints and subject to environmental changes.

Energy is in a headlong rush to the very far-off state of entropic doom. The universe dissipated much of its energy in the first few moments, starting time, creating space, and dissipating energy up to the limits of its early constraints, so now it is building ever-larger dissipative systems to dissipate the rest. It has cranked the engine of order creation for 13.72 billion years, creating some marvelous ways to dissipate energy, and creating a lot of functional order and holosystems in the process.

Functional order is what has resulted from an elaborate sequence of dissipative system formation that makes energy flow faster. That hurried teleology, attributed to energy by the entropy law itself, is the driver of the processes that created, not only the particles of high-energy physics, not just the star systems and galaxies, not just the living surface of the planet earth, and not just the collections of information that reside in genes, minds, and libraries, but all of it. One can view this as if the nested holosystems did it all through their striving, but this is not the whole story. The energy in its flowing has really done it all, created the holosystems and powered their work. Both views are correct, because the holosystems are nothing but energy flowing in successive levels of organization.

As noted above, the general concept of a dissipative structure is too broad to serve as the subject of the theories of Thermodynamic Natural Selection, Universal Evolution and Levolution. Bénard cells, convection cells, swirls at drains, and tornados are discrete entities, but they are not essentially composed of discrete entities.

For this reason, I use the only difference that I know, between these normal dissipative structures and holosystems to distinguish and define the latter. *Holosystems are dissipative structures that are composed of a population of dissipative structures on a lower level of organization.* The parts must be dissipative structures, but they can be, and typically are, also lower level holosystems.

Holosystems capture energy from their environment, generally in a specific form, and they use this energy to do several important things. They use it to create and maintain their internal order. They distribute, and in some cases allocate energy and materials to their component holosystems. They then channel and aggregate the energy output of component holosystems into their own perspective on energy use, which may be internal or external. In its use, and in its entropic losses, the captured and used energy is eventually all dissipated to the environment. Indeed at their eventual demise, even the stored energy of their "body" is all dissipated as rapidly as possible.

It has been a considered strategy on my part, to lump the order of dissipative structures and the order of holosystems together, and call it entropically functional order. The rationale is that both are functional and energy moving models or structures under the drive provided by the entropy mandate. Nevertheless, the structural requirement of part-whole duality adds a considerable difference, which seems to make them more durable. Hurricanes don't evolve, although they do sometimes spawn tornadoes. The ordered energy within holosystems forms the very durable, long standing, and still-building, nested holarchy that is the functionally ordered content of the universe.

Information

The other types of order are connected through and through with this Holarchic structure, but most of them are irrelevant to this story. There is little doubt that the Thermal-Kinetic order represented by ice crystals, for example, is somehow related to the mission of entropy dissipation, but not necessarily by actively moving energy downward as in holosystems. The bonding in crystals and metallic lattices, for example, are the results of heat energy leaving the scene, rather than the result of becoming active agents of its transfer.

There is also little doubt that the type of order represented by information is also related to the mission of entropy dissipation. This type of order seems like an innovation of holosystems at chemical and electrochemical levels, but this type of order encodes the physical, functional order of holosystems. The relevant holosystems to consider with regard to information would seem to be genetically endowed biological ones, on one hand, and humans, with their brains and cultures, on the other.

Since the work of Claude Shannon, the science of information has taken on a life of its own. It even has its own conceptions of entropy which, while it may be related to thermodynamic or energy entropy, is defined in such a way (as uncertainty) that it is clearly not the same thing. It is perhaps too late to define information scientifically in any way other than Shannon's.

Information, however, is clearly a form of order, and in my estimation, it already existed when it popped up in biology. In the case of genetic information, it appears as functional order that has been abstracted chemically from the actual holosystem that it occurs in. The functional order of a biological holosystem is encoded as the information of patterns of DNA.

In the case of ideas, I think the human brain has found a way to mimic environmentally-sensed reality, also in an abstracted space, essentially building a mental model of the world which can be manipulated inside the safety of the skull. In our brains, we decide which connections between associations are real and true, and which are not. A lot of pruning is going on. The result is our individual reality model.

In the case of cultures, the reality models of individual minds are aggregated and manipulated in many arenas of natural selection like politics, academics, and public discourse. Collective and judgmental, idea-selection mechanisms abound in cultures. These systems have actually started to think and create a collective system of beliefs. This reality-reflecting model is similar to what was called the Noosphere by Pierre Tielhard de Chardin, but his conception was more like all the ideas without any selection.

Entropic Drive and the Big Dissipation

I did not go looking for a "prime mover" but in this quest to understand Levolution, I did find one. It is a force of cosmic proportions, backed by a teleological purpose, derived from an unbreakable natural law of thermodynamics. Those are the outstanding credentials behind what could be called the "entropic drive of energy".

The Big Bang would be more accurately described as a Big Dissipation. Energy cannot just sit there as a singularity, according to the observed natural laws of this universe. If a pure-energy singularity was the beginning, the entropy law and the maximum entropy production principle say that it must immediately begin to dissipate, to reduce all gradients, and to do so at the fastest rate possible.

The resulting entropic drive powers everything, informs the profound effects of thermodynamic natural selection, and adapts holosystems to be parts of their wholes. In the beginning, there was no space for energy to dissipate into, and no time to do it, so the first act of the singularity-dominated universe was to dissipate itself and make space at the same time.

This process, as I have come to imagine it, was initially accomplished by primordial electromagnetic charges discharging, without the benefit of electrons, between the two poles of cosmic polarity, between the inner core and outer shell of the singularity. The polarity between positive and negative is derived from the original polarity. The raw electromagnetic discharges degraded into space. The space was required for two reasons. The first is that energy dissipation from the singularity requires space by the definition of dissipation. The second is that entropy required the reduction and obliteration of the strong gradient between the profound opposites of the cosmic poles. Space is the only way to create and increase the distance between these poles to reduce the gradient. The situation is very similar to the act of pulling two strong magnets apart.

Into this new space, a universe's worth of energy would dissipate. The Big Dissipation started creating the expansion of the volume of space and started the clock ticking forward. We know this now as the fabric of our three-dimensional space. We often fail to realize that space contains a phenomenal amount of energy; all the energy of the universe dissipated into it. Gravitational holosystems consume the energy of space, as I will describe later.

Time provides a fourth dimension that began here, but implementing this construct about time from relativity theory results in a static and determined snapshot of our dynamic universe, and that is not helpful to my objective here. I keep time separate so I can talk about energy flowing over the course of time. We ride on the time dimension, and while time is real, it is something we have embedded into our language and worldview. It is possible, but not always necessary to treat time as a dimension. It is perfectly ok to think about space over the course of time. We do not always have to think in terms of spacetime.

The universe inflated due to the rapid creation of space. The dissipated primordial electromagnetic energy was changed and degraded into the so-called "vacuum energy" of space. Energy would have dissipated immediately and fully up to the limit of the universe's constraints on entropy dissipation. Radiation, the energy of photons, did what it could to dissipate the Singularity, but probably within a short time, the primordial energy that could simply dissipate outward in space, and downward to no potential, would have done so.

This leaves the smallish, but now energy- and space-filled universe. It was energy trapped at a high potential, but dammed up in some way by constraints on its faster dissipation. Such constraints on energy are like natural gas fumes lacking a spark. It was going to take a new innovation to tap the pool of energy pervading space, to break down the constraints, and let the energy move along its way toward entropic doom. This innovation in energy flow, I suggest, will create our matter, and us, and the process that began right there, continues today. Indeed, it is very close to being the only thing happening today.

A NEW ENERGETICS

Alfred Lotka (1922) and H.T. Odum (1966) knocked at the door, and so have Brooks and Wiley (1986). Incidentally, some of these luminaries are peripherally a part of my academic training. H.T Odum had been the Director of the University of Texas Marine Science Institute years before I did my Master's degree work there. His legacy in ecosystem energetics was clearly felt, even during my time a few years later. Also noteworthy, Ed Wiley taught my class in "The Evolution of Fishes" at the University of Kansas, and memorably defended evolution against an onslaught by the "intelligent design" camp in a major speaking event there back in 1976.

Rod Swenson will have his day with the Law of Maximum Entropy Production, I believe. This law plus classical thermodynamics, plus dissipative structures, yields a cogent and logical support for everything that I will need to derive Levolution and the Holarchy of Nature.

But some of these are really laws about energy creating order, rather than purely about the laws of energy. This is why I create a break here and recognize Maximum Entropy Production as a Fourth Law of Thermodynamics, but place the Dissipative Structure Law as the first of the Laws of Functional Order.

The complete set of the natural laws of energy and order must include all these, but they seem to belong in different sets. The distinction is that the Maximum Entropy Production Law applies to pathways of energy flowing on its own, not necessarily to discrete quantifiable entities. The Laws of Order are all about discrete, naturally occurring systems, like dissipative structures and holosystems. Thermodynamic

Natural Selection may then logically apply to these, as might quantum theory, and these may make sense because it applies only to discrete entities.

With these laws of energy and of functional order, I can then stand upon their shoulders comfortably, and attempt the improbable feat of inscribing yet another new set of natural laws that take us even further. The situation is that I have found the holosystem model, a derivative of these ideas that leads directly to Thermodynamic Natural Selection, Universal Evolution as a mechanism of general change, and Levolution to explain both the levels of organization, and the resulting Holarchic structure of reality.

Recognition of all these newly codified natural laws will transform an outdated thermodynamics into a revamped science of energetics, and place it at the center of cosmology. The universe is 100% energy, and 100% of the phenomena in it are fueled by energy. Furthermore, 100% of the holosystems are subject to Levolution and Universal Thermodynamic Evolution. These facts have a bearing on the structure of the entire project of science. It is a good thing I am naturally very humble.

Thermodynamics came to Lord Kelvin from industry, the practical, profit-making endeavor of creating power from heat, and it came to live in physics, where it applied broadly. Basic thermodynamics is solid, but it has been rendered incomplete and inconsistent by new discoveries. It has apparently become confusing, as it often leads people to make incorrect statements or interpretations, especially about order.

Even the name of "Thermodynamics" is an anachronism. It is not just about the movement of heat anymore, as is often stated. Heat conduction, molecules in motion, which is what Lord Kelvin had in mind when he coined the term in 1854, is today recognized as the slow way to move energy along its way, toward a final equilibrium. Energy tends to move much faster in dissipative structures. That's why the universe is full of them. They relate directly to the need for speed in the cosmic project to dissipate energy.

Holosystem Energetics

It is easy to envision a new, interdisciplinary science of *holosystem energetics* that keeps the thermodynamic laws generally intact, of course, but adapts them to the understanding that energy in this universe, while it also radiates and conducts, and simply dissipates as kinetic motion here and there, most commonly flows (as fast as it can) within and among holosystems.

Holosystem energetics would add entire sections of new principles about the regularities of energy flow in and through dissipative systems, including how they are created by the flow, and how they evolve and become adapted through the flow. In other words, the subjects of this book may essentially form a new science. Holosystem energetics would study the detailed workings of the simple energy flow pattern that has been repeated at least twenty-three times by Levolution, and has created as many new levels of material organization.

The Holarchy of Nature

Two structural patterns are created by Levolution. One is the pattern of holosystemic, functional order, which is the characteristic pattern of energy flow through a holosystem and its first tier of parts. The second type of structure is a larger scale pattern of several levels of nested holosystem types. It is called a *holarchy*, and it emerges as the big picture of nested whole-part dualities. Figure 5 below, is a depiction of the holarchic structure of the universe.

This latter pattern, though initially conceived somewhat differently, was termed the "Holarchy of Living Nature" by Arthur Koestler in his book The Ghost in the Machine. (1967). It is the nested system pattern that indicates levels of organization among units that are both a whole composed of parts, and a part of a larger wholes. This pattern is how our units, the holosystems, when conceptualized together and connected by whole-part relationships in a series of levels of organization, are structurally arranged. Koester's name for a single whole-part duality unit in this structure, was "holon." It is in the position of the holon that I have inserted the more theoretically dense, holosystem model.

While I am an enthusiastic proponent of the generalized Holarchic structure, and the utility of the whole-part duality concept within Koestler's notion of the holarchy, I do not want to attach my various theses here with the other philosophical contributions of Koestler. The same must be said of another modern day philosopher, Ken Wilber, who has based many of his writings upon it. This is said simply because I intend for this book to contain nothing but serious science. These writers take it further into areas of social philosophy and spirituality, where it loses rigor and seems to be for a different audience.

Holosystems become arranged in a holarchy through the operation of a universal, and dual, relationship pattern. Every holosystem in the universe is both a part-whole duality, and a system-environment duality. Our bodies, for example, are the environment of our cells, as well as the whole composed of those cells. Similarly, we find ourselves in an environment that is an elaborate system of cultural and ecological communities, and we are also parts of these.

The structural pattern of the holarchy among types and levels of holosystems has, rather obviously, been created by Levolution. This process is described in detail in Chapter 6. While sub-atomic particles and gravitational systems still have their mysteries, they have given us plenty of circumstantial evidence that they too are holosystems.

A simple line graph plot of cosmic, functional, holosystemic order over time, yields axes of relationship representing the fundamental forms of energy. Each stems from a particular kind of nested holosystem along lines denoting whole-part relationships, and only rarely to they combine across multiple fundamental forms of energy.

They form line segments connecting holosystems in part-whole relationships that connect along Particulate, Dark Energy, Gravitational, and Electromagnetic Axes.

Figure 5—The Axes of Functional Order

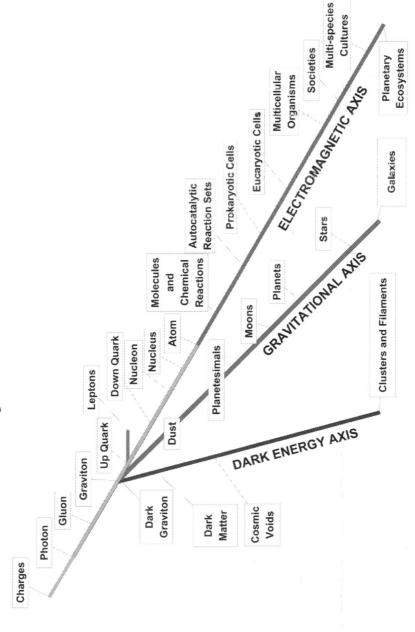

From the likely first dissipative structure, the photon, to the Standard Model particles there is a clear display of levels of organization. The quarks, nucleons, and atomic nuclei are aggregated by residual gravity and the strong force primarily. From nuclei to stable atoms, and all the way through chemistry and the initial levels of biology, all the things ordered by electromagnetic energy or electron bonding form a prominent axis. That includes chemical reactions, families of them, and eventually the cell. From the cell onward, the Electromagnetic Axis holarchy seems to transform into something else, but upon deeper analysis the cultural and ecological systems are, at their base, electromagnetic phenomena.

The Electromagnetic Axis of Functional Order extends all the way through some extraordinary holosystemic patterns of biological, cultural, and ecological order. This branch is more familiar to us than either the Particulate or the Gravitational branches, probably because we are on it. It is still true however, that the seven types of holosystems studied by advanced chemistry, biology, sociology, and ecology are only known from planet earth.

The Universal Functions of Holosystems

While the detail of a given system's holosystemic order is particular to the holosystem type, a characteristic general pattern is actually defined by thermodynamics in the dissipative structure principle, and by the Holarchic arrangement of nested parts. The processes are depicted in Figure 3.

The processes of (1) energy capture from the environment, (2) energy use in many kinds of order maintenance functions, and (3) entropy dissipation to the environment, are mandated by dissipative structure theory.

The processes of (4) manifold distribution of energy to the subject system's holosystem parts, (5) management of combined flows of useful energy provided to the whole by its holosystemic parts, and (6) management of the entropy of parts are mandated by the nested structure itself, or the whole-part duality of holosystems in a holarchy.

This very generalized; "six functions" pattern of energy flow is what is referred to here as both "entropically functional" and as "holosystemic" order. Each real holosystem has its own details, and a complex system's holosystemic order may be simple or complex. It does not really matter. Neither simplicity nor complexity are the goal. Energy flow rate maximization within operational constraints is the goal.

Holosystemic order represents the ordered pathways of energy flow through a dissipative system, a flow that always includes the six universal functions. This flow pattern arises to speed the flow of available energy through the system to the maximum it can handle. Below some energy flow rate threshold, a holosystem ceases to exist.

To maintain their functional order, holosystems face challenges. This is especially true if their environment changes. They must adapt as necessary or disintegrate. Adaptive change always happens the same way, by natural selection acting on a plurality of holosystems, providing differential survival among variants. Whether natural selection acts, as in biology, indirectly on genotypes through the phenotype, or directly on variant holosystems, natural selection is the force of nature that keeps current holosystems <u>adapted</u> to their environment, which within this paradigm is also the whole of which they are a part. This process is Universal Evolution by Thermodynamic Natural Selection.

The holosystem types of the universe have been <u>created</u> by Levolution as sequentially higher levels of organization, as part-whole dualities linked by part-whole relationships. A new type and new level of nested dissipative holosystems always *levolves* as new part-whole dualities in a serially nested holarchic structure. A holarchy is like a hierarchy, but it is sufficiently different to warrant the distinct term despite the fact that many authors have used the word hierarchy to describe the same thing. I find Koestler's term more connotative of wholes and parts and nested structure.

The holarchy often gets a proper name in this book because it is <u>the</u> Holarchy of Nature, a singular thing, a sequential nesting of particular holosystem types in part-whole relationships. The Holarchy of Nature represents the universe's entire apparent content in a single, entropy-powered structure. In essence, it is the Singularity, unrolled by energy dissipation and degradation into the holosystems of nature.

WHAT DO WE KNOW ABOUT HOLOSYSTEMS

Having invented the holosystem construct and seen where it leads, I can now more fully understand what Ludwig von Bertalanffy was trying to accomplish with his General System Theory (GST) in the mid-twentieth century. He spoke of it as a new discipline that would change our perspective. I think it did.

When you can define the fundamental characteristics of your subject generally, but within a tight set of theoretical and thermodynamic constraints, you automatically "know" a lot about your subject. When the generalized system is then "placed" in the habitat of its environmental context you learn even more because all system-environment relationships have commonalities, that I would call ecological. When the subject's environment is then viewed as a "whole" of which the subject is an evolving part, some more facts become known. When the environment is a system that has the same energetic mandate and thermodynamic challenges as the subject itself, more still is learned, and nature gets closer to being understood as it really is.

No matter what type of naturally occurring system you want to talk about, the environment is always infused with the laws of thermodynamics, the mandate to dissipate energy rapidly, the system-environment model, the regularities of ecology, and the regularities of parts in wholes. Then if you look closely, you will see that each of these systems, your subject and its environment, are both just temporary systems in a stream of energy. They are each unique and slightly different, but they only have a lease on life and are individually mortal. Their individual, but differential, deaths will emerge at the population level to look like what we call natural selection operating to keep them adapted as a general type over time. Your subject becomes ever more understandable when it can be viewed as a holosystem.

In the nearly 14 billion years since the Big Bang, most of the energy that could dissipate, or disperse outward by radiation or conduction, or some other kind of rapid entropic degradation, have already done so, at least to the extent that constraints allowed.

In our era, by contrast, there are pools of potential energy stored by constraints, and cycling through death and rebirth as stars which create new forms of matter, and radiate energy as light.

Due to the Second Law of Thermodynamics, the entropy law, there will always be a tendency for these pools to be drained; for the constraints to break down, and for the energy to be dissipated maximally. Holosystems arose as dissipative structures and as the way to do that, and they just kept going. It is all spontaneous. These systems have emerged through Levolution along the Axes of Order, which are the forms of energy, as shown in Figure 5 above.

Increasing the Knowns

Knowing that some entity is able to be generalized as a holosystem, tells us a number of things about it, before we even see it. For example, it says that (1) it is a dissipative structure, (2) it has boundary conditions, (3) it is composed of parts that are also holosystems. Because of this, it means we also know that (4) it arose by Levolution, among aggregates of free-living dissipative structures, and that (5) it remains adapted through thermodynamic evolution. We also know that (6) it does work and (7) such work is observable as a part-role function for a still-larger holosystem that is observable as its environment.

We know that (8) it is a temporary system with a finite lifetime, and that (9) it was encouraged to form by flowing energy, and (10) its mandate is to keep it flowing as fast as possible within constraints. We also know that (11) it captures energy from its environment, and (12) it distributes energy to its parts and (13) uses it internally to create and maintain its holosystemic order. We know (14) it expends energy externally in doing whatever it's "reason for being" may be, and we know (15) it dissipates and degrades energy, as entropy, to its environment, in sufficient quantity to both survive and qualify as a dissipative system.

That is a quick 15 things we potentially know about every holosystem in the universe, and virtually everything in the universe is either a holosystem, or some sort of population of them, or some artifact left by them.

From quark to galaxy, and from atom to ecosystem, we now have a lot more relatives. Like us, they are holosystemic patterns in the flow of energy. They were produced by Levolution, and they remain adapted to their environment through Universal Thermodynamic Evolution. Not only are we now at home in the universe, we are among our kin; sculpted into holosystems by natural selection in accord with a universal, thermodynamic model, the very same thermodynamic model underlying everything else in the universe.

The Levolution-Connected Cosmos

The picture of the universe offered in this book is perhaps not very different than you have already put together from your own experience and learning. If your image of the universe, after reading this, is different than you had pictured before, hopefully one of the main differences is that it seems more unified and connected.

This is due, in part, to the many similarities that exist among its holosystems, and the fact that their motivations are similar as well. They are doing what they can, giving their all, working hard, and striving to move energy downhill, discharging the potentials, using the resources, to get to lower, more dissipated states. The holosystems of the universe all share the same mission of entropy increase. We and they are all essentially in a cosmic alliance to dissipate energy.

The similarities between all holosystems, whether nuclear, gravitational, or electromagnetic, are easy for me to see now, but they were totally unexpected. I did not set out to explain such a striking pattern of unity. We usually advance science by recognizing and appreciating differences and diversity, but the universe is also fundamentally unified, at least in matters pertaining to energy.

Levolution is what connects it all with part-whole relationships all the way up the line. Just as all "life" is shown to be related by the phylogenies of Darwinian evolution, all levels of the Holarchy of Nature are shown to be related by the Axes of Functional Order produced by Levolution.

Holosystem Universals

Initially, I defined holosystems based on the similarities observed to exist among a special class of dissipative structures. It was a theoretical construct that I had invented by synthesizing several solid notions of generalized system principles. It arose because it was needed to go beyond the dissipative structure in summarizing the similarities observed in all the natural systems. In other words, I took it upon myself to expand the basic concept of dissipative structures, which seems entirely thermodynamic in nature, to account for the fact that the particular ones that I wanted to talk about were composed of similarly structured parts.

However, I am now of the opinion that the holosystem construct is actually a real and notable thermodynamic structure in the same way that dissipative structures are notable thermodynamic structures. The holosystem construct is a sound model that is easily defended, just by reviewing what we already know about real systems in nature.

The model is also widely applicable, and easily understandable, based as it is, mainly upon common observations. It has already proven (to me) to be a helpful construct, and a central one, in understanding how the universe works. I now believe that holosystems and holosystemic, or functional order are real features of energetic reality. I believe that the shared ideal form or model, used in common among both parts and wholes, is essentially what allows Levolution to act.

Actual holosystems, upon examination, must fit the principles involved in the supporting theories very well, or they must be excluded. Organs and organ systems, for example, are excluded from the list of holosystems, because they were never free-living entities, even though they look like the important parts of multi-cellular holosystems. Even with this kind of discipline, we find that the construct is broadly applicable to the phenomena of most of the known, naturally occurring systems in the universe. Rocks, minerals, tissue types, organs, etc. may be aggregations of various holosystems, or populations of them, which we have named, but they are not holosystems as I have defined them. Particles, atoms, stars, and cells, however, do fit the bill.

This state of affairs, the detail and explanatory power of the holosystem construct, coupled with the observed broad ability to match it to the existing natural and cosmic systems, is screaming out for us not to let the similarities here be dismissed too lightly. My sense is that this is a very important construct indeed.

Holosystems are well-defined by the (1) thermodynamic, (2) systemic, and (3) holarchic structural similarities they share, but their existence in a naturally selective environment, and their similarity of entropic purpose, add to the similarities. By the time a holosystem is defined in general, the definition applies equally well to quarks as it does to ecosystems and galaxies.

Holarchic structure, basic energetics, and functional similarities define the holosystem construct, but these may not exhaust all of the important similarities among them. In addition to being inherently similar in the above three ways, and sharing of a similar teleology, or purpose, they also face similar situations in their environments. There could be, and I think there is, a generalized *holosystem ecology*, which goes even deeper.

Holosystem Ecology Unveiled

There are survival-related similarities in the challenges that holosystems face in their environment and in their solutions to them. Environments constrain existence as a holosystem, and they naturally prevent or select out the less fit holosystem attempts because energy takes the fastest pathways available, even when the pathways are discrete holosystems. Natural selection is a profound truth, and it can be seen as a similarity of action by their "containing environmental wholes" that every holosystem experiences.

As I must keep noting, natural selection is what physicists and chemists would normally call the action of constraints. Many environmental similarities stem from the system-environment model, which is well-developed in the science of ecology, but new ones stem from the part-whole duality model suggested by Levolution.

These principles may also be generalized easily to the other holosystems. The maintenance of thermodynamic stability at the top of the "dome of attraction," a steady state defining the fastest energy-flowing condition possible, is a concept of dissipative systems that captures the fundamental dynamics of a structure itself. But whether viewed as such a system in its environment, or as a part within a larger whole, these are both easy templates to apply to nature's systems. I think we will need to adjust ecology to emphasize the latter view in due time.

Teleological Similarities

The entropic teleology of holosystems, their very purpose, is something shared in common by all of them. It does not really matter whether energy in the universe is allowed by Francis Bacon or proper science to have a "purpose." Energy will still act as if it does. Goals, purposes, final causes, are not reserved for mentally guided processes. The entropy law gives energy a direction and a purpose. Dissipative structures have inherited this, but the purpose is modified a bit by them, because of the relationship between being (simple existence) and not being. If a structure is not there, it will not move energy, so each one of them might be said to inherently seek survival, whether conscious or not.

Minds are not the only place where final causes, goals, specific end-states, or specific directions can be mandated by natural law. Thermodynamic systems like dissipative structures and holosystems, have the purpose of increasing energy dissipation and degradation built right into their very existence. Dissipative structures and so holosystems, take the entropic mandate or teleology and turn it into a tendency to seek their own survival (so they can be there to dissipate more energy).

There may also be a tendency among some mindless holosystems to seek energy efficiency so a greater plurality can maximize the use of available energy as it gets scarce, or so that a given plurality can increase the time of their survival.

In other words, survival itself becomes a holosystemic goal well understood in biology, but I would bet that if stars, for example, could think and talk they would prefer existence while they exist. Among animals and cultures, their functional order may survive their death because their functional order is encoded in genes and ideas.

Similarly, while the functional order that makes a star is not encoded in genes, or the laws of chemistry, it is encoded in the laws of gravity. This holosystemic order will exist even after they materially, as individuals, explode or die, as long as gravity operates.

CHAPTER 5

THE LAWS OF ENERGY AND ORDER

All things are an exchange for
Fire, and Fire for all things, even
as wares for gold and gold for
wares.

– Heraclitus

THE LAWS OF ENERGY

Thermodynamics is a misnamed set of antiquated and incomplete, but very fundamental natural laws that reflect the observed regularities in the properties of energy. Because the universe is only energy, and because dissipative structures, and their cousins, the holosystems, are mere patterns of energy flow, these laws are quite important. Here we will review classical thermodynamics briefly, and then spend more time on a recently proposed law of energy.

Then I will continue and take the necessary liberties to propose six more natural laws, the "Laws of Functional Order". All of these are essentially Laws of Energy, but the two sections of laws covering energy and functional order respectively, are descriptive of energy when it flows at large, and energy flowing in ordered structures, respectively. Only the first three of the ten laws here are well accepted, but the proposed fourth law of thermodynamics, the MEPL, is likely to be soon, as it is getting some corroboration in the literature, but the last six proposed laws, which are the Laws of Functional Order, are only here and now set forth, for thinking people to examine for themselves.

Law 1 – The Energy Conservation Law

The first law of thermodynamics states that energy cannot be created nor destroyed. The amount of energy represented by the Singularity at the beginning of the "big bang" is the big "given" in the classical science of the universe, and the energy conservation law says that we will not be getting any more than the universe already has. It also means that energy must always be completely accounted for, in any energy transactions that we hope to fully analyze, including mere changes in form.

Energy can become matter, and matter can become energy, but energy cannot be created or destroyed. Matter is related to energy by a simple constant of proportionality, as Einstein showed us, and that constant is the speed of light squared.

Law 2 – The Entropy Law

Classical thermodynamics is mostly simple and intuitive, but the principle of entropy, in particular, is so pervasive that people in various fields and situations have overlain the basic principle, that energy always flows to lower states in terms of levels of energy potential, with less fundamental meanings that may not always be true.

For our purposes we need only the fundamental understanding, but it should be noted that the confusion nearly always relates to order, isolated systems vs. open systems, or constraints on energy's flow.

The *Second Law of Thermodynamics, or the Entropy Law,* says that the quality of energy called "entropy" must always increase, which simply means that energy itself experiences continual dissipation and degradation in everything it does.

Energy is always driven by this law to lose its potential, to disperse, dissipate, or spread out. Energy can change form, but not without a cost, and spontaneous changes, those which happen in energy's natural course, are always toward a more degraded form.

Energy can be stored, but only temporarily, and never without a loss to entropy. Energy generally moves outward in space from the Singularity, forward in time, and downward in energy potential. It is always becoming more spread out, and less concentrated. Time never goes backwards, even though quantum physicists do not seem to care.

Human technologies are often interested in concentrating energy, and increasing the work potential and power of machines. These things can be done by thermodynamically understood systems, but there is always an energy price to be paid, more energy must be supplied from outside, for going in that direction, which is upward in potential. On a net basis, considering also the input energy, the entropy law is always observed. For energy in general, getting to a higher potential is going uphill, so work must be done to do it and also account for the entropy law. Systems can get there, but only by expending more energy, such that the second law is not violated in an overall sense.

Law 3 – The Absolute Zero Law

The *Third Law of Thermodynamics* states the impossibility of ever getting something so completely lacking in kinetic energy, so cold and motionless that it reaches the temperature called Absolute Zero, as measured in degrees Kelvin. Zero on the Kelvin scale is equivalent to about -273 degrees Centigrade. This essentially means that one cannot find a place where there is no kinetic energy at all. No matter what, where, or when, there is some energy everywhere in the universe. This law is about temperature, which is a measure of kinetic energy, so this particular law is specific to that form of energy. It does not say anything about the "rest energy" held in the form of some particle. It refers only to the motion of the particle. However, even in the absence of matter, there is also no way known to evacuate all the energy from space, which as we will see, is relevant to our tale.

The Equilibrium Principle

Some authors have seen fit to include this principle of thermodynamic systems as a thermodynamic law. It is often called the Zeroth Law. It states that given three systems A, B, and C. If systems A and B are in thermodynamic equilibrium with each other, and system C is in equilibrium with system A, then C is in equilibrium with system B also.

All I would note about this principle is that it seems like a rule of pure mathematics, and because it is about energy systems in equilibrium and not about energy itself, I would move it to the developing Laws of Order, which are here currently limited to laws related to non-equilibrium thermodynamics, the Laws of Functional Order, based on dissipative structures. It is, at any rate, not relevant to our discussion. In all cases, we will be discussing systems far from equilibrium.

It is worth noting here that equilibrium, the state of no energy flow at all, is a state at the very bottom of what is known as a "basin of attraction" in terms of its mathematical description. Energy flowing at low levels of intensity, and given no outside energy with which to build order will tend toward equilibrium. Equilibrium is entropic doom, and the lack of any kind of order.

Remarkably, the dissipative structure is also a state attractor in a dynamical sense, but it is quite different. This attractive state is the point at the very top of a "dome of attraction". Flows of energy with increasing intensity, are allowed to move faster, in the direction away from equilibrium, by patterns of functional order as they develop.

As the pattern develops toward the dissipative structure state, the constraints increase to push it toward the universal ideal pattern of the dissipative structure or holosystem, where entropy production is maximized. The driver here is natural selection. Some people would call this state, on top of the dome of attraction, a peak in a fitness landscape.

The Reciprocal Relations Principle

In thermodynamic exchanges that involve material exchanges as well, there are reciprocal exchanges of materials that balance in the transaction as reciprocal ratios, just as energy quantities balance due to the conservation law. This is usually known as Onsager's Reciprocal Relations Principle. It has some important ramification in certain areas, but I have not used it directly in my thesis here.

Law 4 – The Proposed Law of Maximum Entropy Production MEP)

The proposed *Fourth Law of Thermodynamics* is new, and has been proposed and well-explained by Rod Swenson (1988) and S.N. Salthe (2010). It is similar to the Maximum Power Principle of ecologist, H.T. Odum, but power is not the same as entropy and so it is Swenson's concept that has won my acceptance, while I do understand Odum's very similar notions about it. The MEP is currently coming on strong in some scientific circles, but is only slowly winning the acceptance of the physics community.

The Maximum Entropy Production Law simply states that the downward flow of energy (prescribed by the Entropy Law) will always follow the pathway(s) that result in the downward, degradive, dissipative flow of the greatest amount of energy in the least amount of time, given the constraints. It says that time is of the essence, and that energy flow rate is important. Fundamentally it is thus a law that dictates that energy will "select" the fastest possible pathways and allocate itself to them accordingly.

The best way to visualize this law in action is to ponder the pathways of water, flowing and falling down a mountainside. I am paraphrasing this from having read Swenson. The topography represents an irregular landscape of constraints, and the water under the influence of gravity represents energy flowing downward. The water is undoubtedly going to "find its level" and in doing so will be found in the steepest and largest channels available. There is nothing to prevent topographic constraints from trapping and ponding up some water at some high potential, but there is also no delay or confusion about where it will flow when it can. By this law, energy selects the pathway(s) that will transmit energy the fastest within applicable constraints, and allocates itself to those pathways.

This law is, very significantly, a law of the selection and allocation of energy to alternative energy pathways. Swenson and a growing number of others believe that this makes it equivalent to the profound mechanism of natural selection. Although I am sympathetic, I disagree with that point, but only on a technicality. I recognize its significance, but we need to get the natural laws just right.

Energy pathways are not always discrete. Natural selection must involve the differential survival of discrete, and variable, entities within a set, group, or population of them. When applied to energy pathways alone, this law is always correct and should be recognized as a Law of Thermodynamics, but only if the pathways were among discrete entities would it become natural selection.

It is important to get the Law of Thermodynamic Natural Selection stated correctly, and unassailably. In my own view, and consistent with my education in evolutionary ecology, natural selection is really an emergent phenomenon and can only apply to populations of discrete holosystems. We will fix that situation below, such that natural selection gets its due as a natural law related to the phenomena of functional order, and where discrete structures and systems are actually the subject.

THE LAWS OF FUNCTIONAL ORDER

We now come to a new set of laws related to energy that are proposed here, for the first time, but they will be familiar. In this separate set of laws, we are specifically concerned with the laws of energy flowing in entropy law compliant, spontaneously-ordered, naturally occurring systems. The above set of thermodynamic laws pertain to energy flowing at large, or in general pathways, as opposed to energy flowing in and among discrete, naturally occurring structures or systems. This distinction is important, because discreteness allows for populations, among which emergent phenomena may occur and be observed.

Structures and systems are overt controls or qualitative constraints on the flow of energy, and as we shall see, they exist because they increase the throughput of energy, not because they constrain it. It is my bold contention that the laws below, which begin with the Dissipative Structure Law, provide the only way known to science that our subject type of order; functional, system-enabling and energy-dissipating order, may be formed in the universe.

Up until now "order" has been a very general, and barely scientific, term, but it is worthy of becoming a very meaningful scientific concept. I hope to help resolve a long-standing confusion about order that has stemmed from the fact that there are multiple types of it.

Reorganizing Thermodynamics

I am not blind to the historical and scientific momentum behind our current arrangement of the regularities that science has found about energy and codified into the current three Laws of Thermodynamics. It is also evident that we have stumbled upon some other important regularities that need to be fitted into it. In the process of writing this book, I have tried several ways to diplomatically, but logically, re-arrange the subject, but adding full-fledged thermodynamic laws seems perhaps too radical.

After living with what I called the Laws of Functional Order for a while, and considered where the Maximum Entropy Production Law should go, I have also come to the realization that there are also sets of Laws regarding the other kinds of order. As I mentioned in the introductory chapter, the Dissipative Structure Law relates to entropically functional order, and a lot of confusion surrounds the fact that Boltzmann was pondering another type, which I call thermal-kinetic order, the non-entropically functional, statistical order of mere spatial distribution.

Order in general has always been a problem because it lacks a consensus on its scientific meaning. There is a very large difference, you will agree, between particles improbably clustered in the corner of a box, and a bird on the wing. Both relate to the seemingly improbable, but the bird's order is not merely positional and allows it to fly while dissipating energy. The particles in the box dissipate their energy by the virtue of the thermal kinetics of heat, but the order that is decreasing in the box is not functional order, it is merely an improbable arrangement of particles in space.

It is now clear to me that mankind happened to discover the science of thermal-kinetic order first, along with the more general concept of entropy, but we are now faced with another kind of order; entropically functional order, which is of a completely different character because it is order that actually increases entropy production.

Dissipative structure formation is ultimately the only known source of increasing functional order in the universe; even the holosystems are all dissipative structures. The increase in energy flow (and therefore entropy production) that is allowed by the order of such systems is also their entropic purpose; their function.

The only difference in fact, is that dissipative structures that are not also composed of parts that are either dissipative structures or holosystems have not arrived here through the integrative process of Levolution, while all of the holosystems have. This is why holosystems have developed in the universe as entities on easily identifiable levels of organization. This is also why the so-called property of "complexity," which I view as simply a compounding of the simple holosystemic order, is generally found among holosystems exclusively.

There is thus proposed a whole new section of the Laws of Energy, specifically about energy operating where it mainly lives, flowing within nested, and ordered, dissipative systems and holosystems. These are the six Laws of Functional Order.

The concept of a "system" in thermodynamics has always been very basic, and its origin was in classical mechanics. Historically, there was recognition of *closed systems* and *open systems*, with the latter able to exchange energy and matter with its environment. This was part of the attempt to reconcile the problem of biological phenomena going against the entropy law, by distinguishing those cases where environmentally sourced energy could be added to an open system, but not to a closed system.

Closed systems in classical mechanics became *isolated systems* in thermodynamics. The situation is much the same as that of *reversible processes,* which similarly cannot possibly exist in nature.

We have now come around to the understanding that there are no systems that are truly closed or isolated from energy flow, except hypothetically. Open systems concepts, based on systems through which energy flows, have now been merged into the theory of dissipative structures. Ilya Prigogine's theory of Dissipative Structures is proposed here as a formal law. It is the First Law of Functional Order.

Law of Functional Order #1 - The Law of Dissipative Structures

This law simply states that flowing energy in a given region may spon taneously create local order in the form of a pattern in energy flow comprising a *dissipative structure*. The pattern increases the overall energy dissipation of the system by capturing and using environmentally-sourced energy to maintain the system's order, and pay the entropic cost of it. The development of such structures is spontaneous in the sense that the overall result is a region that moves energy downward in potential faster than it did before.

The dissipative structure itself, the pattern of the energy flow pathways involving the (even more fundamental) requisites of external energy capture and external entropy dissipation, allow the internal segments of the energy flow to do the work cycles that are the requisite functions of such systems, including creation of its internal order. The order of the structure, whatever its details might be, always speeds energy flow in comparison to the state or situation before it formed. The acceleration of the energy flow is why they form in the first place.

Dissipative structures are essentially an innovation in energy transfer and entropy production, and it compares, both functionally and favorably, with other energy-moving mechanisms like conduction, diffusion, simple emission, or radiation. In the creation of dissipative structures, energy creates order in the form of a new discrete entity, while it achieves faster overall flow. This, of course, better serves the mandate of the entropy law to dissipate energy, and the mandate of the new Law 4, which is to do so as fast as possible.

Dissipative structures capture energy from their environment, use some of the energy to create and maintain their order, and in the overall, dissipate more energy than was being dissipated before they existed.

Formation of a dissipative structure is the only way known to science that any kind of functional, systemic, energy-transferring order, which I here call "functional order" or "holosystemic order" can be naturally produced. Outside of dissipative structures, order is simply obliterated by entropy.

Tornados, hurricanes, convection cells, and the spiral flow at your bathtub drain, are dissipative structures. You and I are also dissipative structures, but we also qualify as holosystems, and these have a special set of applicable thermodynamic principles. Dissipative structures of all kinds, however, are the universe's way of speeding energy flow downward in potential. They are very common, and are often associated with hydrodynamic phenomena because here we can see energy flow patterns (usually thermal or gravitational) as material fluid flow patterns.

Functional order can spontaneously increase in the universe, as long as it increases in the form of a dissipative structure. This little-known principle is a supremely important cosmological and thermodynamic regularity. Dissipative structures must include gravitational systems because we observe gravitational order. The cosmic order that astronomers see out there – all the spheres, orbits, and radially symmetrical objects, represent the gravitational ordering done by dissipative gravitational structures. As we shall see, however, viewing gravitational systems as energy-hungry holosystems will both require and yield many new perspectives.

We now have a new, young, and growing set of thermodynamic principles that begins with dissipative structures, and I will explain how this both began and continues the story of increasing order in the cosmos. The spontaneous formation of thermodynamically-created order in the form of dissipative structures becomes the first Law of Order, but this is only the beginning.

The Dissipative Structure Law is a logical extension of thermodynamics, it is worthy of elevation to the status of thermodynamic law, and it is the obvious foundation for the many other principles that we will see apply to such energy-transferring systems as they go on to create the order of the cosmos. A section of thermodynamic law to address dissipative structures, holosystems, and the specific energy-related regularities that characterize such systems is what has been missing most acutely from the project of science. It has essentially been missing from physics and cosmology, and every subject based on these. This is the best argument I have for using the word paradigm to describe the upcoming perspective-shifting arguments in favor of Levolution.

The Law of Maximum Entropy Production (proposed Law 4) applies to energy at large, but it also has an important role in the thermodynamics of ordered systems. When applied to discrete holosystems, as it will be below, this "selection of the fastest pathways" principle will explain this same phenomenon as the root cause of holosystem formation and termination, which on biological levels is interpreted as life and death. This individual level phenomenon of simple existence will then "emerge" in populations of holosystems to become a major component of the rationale for Thermodynamic Natural Selection.

The focus upon ordered systems in this section of energy-related laws is what allows us to see the reality of natural selection as a very fundamental thermodynamic principle. It also provides a logical place to codify the natural, energy-driven processes that have only now been discovered to be fundamentally thermodynamic in character. Holosystems as discrete dissipative subjects and a principle of Thermodynamic Natural Selection, under which they operate, is a framework that biologists will immediately recognize. These principles will become the basis for a universal and thermodynamic version of evolution. These principles all relate to holosystems only, but they are thermodynamic in origin, and derivation. They are all about energy dissipation and degradation by naturally occurring, ordered systems, and they explain virtually everything.

Ilya Prigogine outlined the fundamental characteristics of dissipative structures, but a number of important regularities about them are worthy of re-iteration. The following principles are what I understand to be the most important concepts to be drawn from the definition of these structures.

The Spontaneous Formation Principle

Dissipative systems, when they form, form spontaneously in increasing energy environments, creating an entropically functional type of order among the flows of energy within a range of energy flow rates. In thermally ordered structures, for example, their formation occurs when the region is heated to a given threshold temperature.

Spontaneous Termination Principle

If a dissipative structure leaves its habitable range of energy flow rates, or for any other reason, fails to perform the required energy functions of a dissipative structure, it will terminate spontaneously, releasing its energy to the environment.

The Functional Order Principle

The entropically functional order of a dissipative structure is a pattern in its defining energy flow pathways that increases the overall rate of energy flow through the structure, which is also a measure its total entropy production. This consistency with the second and fourth laws of thermodynamics is the key reason for the existence of these structures.

Environmental Energy Capture Principle

A dissipative structure must capture or absorb energy from its environment in sufficient quantity to fill all of its energy needs, including the energy costs of capturing more energy, internal order production and maintenance, external work, and entropic energy dissipation or degradation.

Environmental Energy Dissipation Principle

A dissipative structure dissipates entropy to its environment. Even the energy used by the system is entropically degraded or dissipated and eventually contributes to the total entropy production of the system, which must be expelled beyond its boundary condition. It is an important point that entropy may be sometimes considered as essentially a value judgment about the usefulness of energy by one type of dissipative structure. The entropic waste energy of one type may be captured and deemed useful by another type of dissipative structure. It will truly have been degraded or dissipated, but it still exists, and may be another dissipative structure's resource.

Law of Functional Order #2 - The Law of Thermodynamic Natural Selection

The Law of Maximum Entropy Production is applicable to energy at large, and pertains not only to generalized energy pathways but also to dissipative structures and holosystems, but it manifests somewhat differently according to the perspectives of such systems. The energy in the universe always maximizes entropy production, within constraints, whether flowing in a system or not. The Fourth Law says that energy allocates itself to the downward flowing pathways that can move the most energy.

When a set or population of variant dissipative structures represent the available or possible downward energy flow pathways, an iron law of Thermodynamic Natural Selection operates to favor, or allocate more energy to, those varieties of dissipative structures that generally maximize the entropy production. This action enhances the survival probabilities of that variant type's survival into the future.

Natural selection does not act upon energy flow itself, but it can affect discrete things like individual energy flow pathways, dissipative structures, or holosystems. Natural selection, like the kinetic variable of pressure, is an emergent phenomenon that only operates in a population of discrete things. In this case it operates to increase the probability that some favored type or variant of a system will be represented in the future to accept the energy flow. For individual systems, its action may be simply termination, destabilization, death, or reduced reproductive success.

A population level perspective is required for generic "system termination" to become natural selection. Natural selection also does not work if there is complete homogeneity among the holosystem population. Variation among discrete entities is an important requirement for its operation. Variation, however, is not difficult to produce. Most variations from existing holosystem forms are not improvements because what exists has already been selected favorably, at least once, and to some degree. However, the conditions of their formation may change, different environments may be encountered, differing forms of energy may be exploited, and variation is thus a very common response to these environmental changes. The vagaries of existence ensure that variation exists on nearly every level of reality.

The biologists are reading intently now. They know that the source of variation is extremely important to the operation of natural selection in biology. In biology it is thought that random, heritable, genetic mutation results in the fodder for natural selection to operate upon. Heritability is important in biology because its subjects are programmed to reproduce and die. They live a pre-ordained life cycle, often of fixed duration. Heritability is not so important outside of biology, where reproduction is not required. Continuity of the type is the key concept here. Protons can probably ignore heritability.

Another point is needed about reproduction in biology to enlarge the scope of natural selection beyond these levels of organization. Reproduction is a two-pronged innovation related to the myriad of chemical molecules that need to be obtained and delivered at just the right time and place to conduct biological life, and perhaps to the reality that biological systems cannot maintain their functional order forever. Elementary particles, even the most fundamental of them, play a different game. The conditions of their creation may continue to exist over billions of years, and they may find stability that lasts for a similarly long time. Natural selection among the particles is what physicists refer to as the constraints of an environment which either do or do not foster stability and survival.

So, we now have natural selection as one of the thermodynamic laws pertaining to order. The logic of this is simple once we accept the Maximum Entropy Production mandate. The logic then becomes the same as that used by Alfred Lotka when he suggested in 1922 that natural selection was indeed a physical principle of thermodynamics. Physicists or biologists should have picked up on this and run with it, but they didn't.

Dissipative structures, as we know, are thermodynamic constructs that are essentially patterns in the flow that can form spontaneously when a threshold rate of energy transfer is reached in a thermodynamic system. When energy or fluid-dynamic flows can be organized, as in a whirlpool at a drain, it makes the overall flow faster. Order is allowed to develop by the entropy law, but only if it results in more and faster entropy dissipation overall than existed before the order was achieved. Order is what is needed to organize the flows.

The new thermodynamic principle that leads to the thermodynamic nature of natural selection is simply this: Just as dissipative structures form spontaneously under certain conditions to dissipate energy to the environment faster, they also disintegrate spontaneously when conditions are not right.

When the energy flow rate slows beneath a certain threshold, when one of the universal functions cannot be performed, or when energy has found a faster alternative pathway, dissipative structures spontaneously disintegrate, taking their order with them. This is the important, but long missing fact that leads to natural selection among the dissipative structures known as holosystems. We can see now that natural selection, Charles Darwin's mechanism for biological evolution, really is nothing but the emergent effect of the individual termination and differential survival in a population of variant holosystems. At stake is the matter of continued survival as a maximally flowing energetic holosystem type. Being such is mainly a matter, as we shall see below, of being the right part for a given whole.

Introducing Thermodynamic Fitness

The converse of natural selection is fitness. The avoidance of natural selection is the thermodynamic sub-goal that helps to explain change among holosystems. The characteristics that convey fitness are solely dictated by the environment(s) of the holosystem. In our new perspective, this environment may be the inside of a larger whole, the perception of which requires something like a particular synapse or circuit in the human brain that one can almost feel activating as we let our attention span upward to a new level of organization.

Among living holosystems, fitness is gained through a good match between the characteristics of the holosystem and the challenges of its environment. The environmental challenges correspond to the universal energy functions of holosystems enumerated above. The surprising, if not shocking, fact is that these challenges, viewed thermodynamically are always the same. While all the details are different between any two holosystems, all the generalities are the same ones. Holosystems on all levels of organization are thermodynamically self-similar.

This leads to a notion that even I resisted until it clobbered me over the head. Fitness is the ideal of being a perfect holosystem in some environment. Selection pressure, the enforcement of thermodynamic fitness, may vary between mild and extreme, but generally, survival depends upon being a performing holosystem in the circumstances encountered. This is fundamentally the case, but it means that we will need to rewrite the ecology textbooks, and improve our evolutionary view of organisms and ecosystems, as parts and the whole of which they are part.

Natural selection sculpts the energy of the universe into holosystems that perform in its various environments to move energy down its proverbial hill. Holosystems are apparently the ideal way in which energy builds up the scale of flows and moves energy as rapidly as possible downward in potential. Plato, if he were here, would have no problem with this, but as an ecologist I can say that it is a tough pill to swallow, because Western science has rejected Plato's notion of ideal forms. The logic, however, is almost inescapable here.

Law of Functional Order #3 - The Law of Universal Evolution

Thermodynamic Natural Selection is the primary mechanism in causing population level changes in holosystem characteristics. Holosystems change by means of Thermodynamic Natural Selection. Another word for this change and differentiation of forms is evolution.

Darwin was conservative in his approach to evolution by natural selection and the impending revolution he knew it would set off. He sat on his theory for 20 years before being prodded into publishing it by Wallace, who was about to do so himself. When he did publish, Darwin was again very conservative. He limited it to biological realms, even though there was some evidence that it extended to other realms, especially the cultural one. Herbert Spencer, Darwin's contemporary, should be noted in this context, but even good ideas can be guilty by association, and association with loose and early interpretations of Social Darwinism, eugenics, and Hitler, were enough to stop ideas of cultural evolution in their tracks until recently.

Today we can see that even Darwin was dealing with thermodynamics when he was dealing with natural selection among his finches. Today we can see that, because thermodynamics is universal, natural selection is universal. Because natural selection is universal, evolution by natural selection is probably happening everywhere. We can note a few interesting examples from outside of biology.

If a discrete system is dynamically stable, it continues to exist. If it becomes unstable, it disintegrates to its stable parts and ceases to exist. So sub-atomic particles that are thermodynamically stable exist, and when they are not, they don't. Thermodynamic stability is a matter of natural selection, and the sub-atomic particles are directly at the mercy of energy fields within which they either do or do not find stability. Different particles are fit in differing environments. Particles evolve by natural selection. The important variation is provided by the differing fields and environments. A population of particles can change in response to their environment, but the change can also be said to be accomplished by natural selection because it is differential survival, differential stability, differential existence, that does it.

If a few reagents are placed in a beaker, a new environment has been created. What we call a reaction is a process by which natural selection operates in the beaker. The molecules that are stable in the new environment are the survivors. Some reagents may not be stable at all in the new environment, and may no longer exist. Other molecules in the soup may find all that they need to combine into a larger system, a new and larger holosystem, a composite chemical molecule. They bond using their valence electrons, but whatever results will, of course, be subject to Thermodynamic Natural Selection. Chemical evolution and apparently Levolution too, can occur in a beaker.

Consider a patch of outer space into which we introduce a few billion particles of differing masses and differing velocities. Will this region of space provide an example of natural selection? Yes it will. The particles will gravitationally clump together in patterns that are dictated by their mass, velocity, position and direction of travel. Some particles will eventually collapse to form a central mass of formerly independent particles. Other particles may order themselves into stable orbits around such clumps, or they may fall into and collide with the central mass. Others may fly off in some random direction; having a speed in excess of the escape velocity of the central clump. These exit the frame of reference altogether. It's gravitational natural selection, where the fitness parameters are Newtonian.

The particles that have lost their independent existence within the subject frame of reference, either through collision or escape, no longer exist as independent particles. They were subjected to *gravitational natural selection*. The orbiting particles survived. The central mass is no longer a particle, but the particles have become parts of a higher-order gravitational structure, powered by the combined gravity of all of its particulate parts acting from a single center of gravity. Both evolution and Levolution have occurred here, and they have occurred by means of Thermodynamic Natural Selection. This concept of gravitational natural selection is very important and will yield valuable insights in cosmology later on.

Law of Functional Order #4 - The Law of Levolution

Thermodynamic Natural Selection may adapt and **differentiate** as in universal evolution, or the same mechanism may **integrate** what was a population of independent, whole dissipative structures or whole holosystems existing independently into the complementary parts of a larger version of a holosystem. The key to this process is performance of the universal functions of this higher-order holosystem. Thermodynamic isonomy, or equality under the laws of thermodynamics, and the dissipative and holosystem structure laws in particular, provide a universal target for natural selection to operate upon in building a new level of organization.

Where evolution differentiates holosystems, levolution integrates holosystems by improving the performance of the universal functions to move more energy downhill. Even if the original population of holosystems is homogeneous, all of the universal functions are already being performed by each individual. When looked at collectively however, the population has some individuals out near the edge, some are near the inevitable entropic waste piles, and some are trapped in the middle where resources are consumed before they can arrive.

The homogeneous holosystems are seen to be in diverse environments. Their situations are automatically different. Naturally, they may begin to differentiate by evolution from these factors alone. The holosystem's functional imperatives, however, always operate. The population of holosystems does not differentiate randomly, it differentiates in *particular directions* that turn the individuals into the complementary parts of a larger holosystem. The new holosystem snaps into view as the above referenced "synapse" activates, and we suddenly see a new, higher level of organization for the first time. Like ecosystems, they may have existed for a long time before we could see them as such. The information you are receiving will help you remove your scales or blinkers and keep that synapse active.

This profound process is called Levolution because it results in a new level of organization in addition to a new holosystem type. Here I am stating the general principle of Levolution, but more details of this important process, which is the central topic of this book, will come. Levolution should be suspected as a cause whenever evolution seems to result in a new level of organization.

Needless to say perhaps, Levolution is the process that many authors have, for years, called "self-organization". Environments evolve holosystems in to parts of a larger holosystem and the integration of these results in Levolution. The holosystem, a thermodynamic form or structure, clearly does not self-organize. Not even biological development, as from a fertilized egg, is self-organization. Outside influences are easy to spot. Prigogine used the concept and that may be the source of the confusion, but look at the source from which the functional or holosystemic order came. To call biological development "self-organization" degrades the essential role of its parents and ancestors, not to mention the energy it consumes from its environment, and the action of Thermodynamic Natural Selection. Self-organization is a concept whose time has gone.

Law of Functional Order #5 - The Law of Holosystems

Holosystems are entities derived from dissipative structures, and I need to explain why holosystems do not appear in these Laws of Functional Order as a lower-numbered entry. The reason is because holosystems require Thermodynamic Natural Selection, Universal Thermodynamic Evolution, and Levolution to arrive at their structure. Holosystems are a kind of culmination of all these laws of change and order creation.

Holosystems are a subset of the dissipative structures. Holosystems exist as a special class of dissipative structures that are composed of parts that are also dissipative structures. They are also discrete, bounded, and distinct from their environment. They conduct the three energy related functions of dissipative structures, but they must also manage their parts, which are doing the same three things in some other way. Their environment is typically the inside of another holosystem, of which it is a part. Holosystems, unlike other dissipative structures, are always found in a nested pattern.

The principles below apply to this special subset of dissipative structures, the holosystems, which are naturally occurring and defined by the fact that their parts are also dissipative structures or smaller holosystems. With regard to their thermodynamic aspects, they are dissipative structures with additional constraints which relate to their self-similar parts.

The theory here is simply a reflection of the observed reality of the holosystem structure. The points below, which define holosystems, are derived from three historic sources; General System Theory, Dissipative Structure Theory, and what I will call Levolutionary Holarchy Theory.

Holosystems are Dissipative Structures

Holosystems are essentially thermodynamic dissipative structures, but they are further qualified and must have parts that are also dissipative structures. From their dissipative structure features, Holosystems get their reason for being. They exist to serve as dissipative structures to move energy downward in potential at a faster rate than in their absence. These are essentially the concepts of Ilya Prigogine.

Holosystems are General Systems

From a General System Theory (GST) perspective, holosystems have boundary conditions, are distinct systems with composed of functional and interrelated parts, and exist through relations with their environment. These are derived generally from the work of Ludwig von Bertalanffy.

Holosystems Are Parts of the Holarchy

In the Levolutionary perspective, a subject holosystem is created from a set of dissipative structures, or lower level holosystems that as a type, were once "free living" or independent wholes, but which have been further adapted (evolved as a population by natural selection) into the forms of the parts of the larger, containing, subject holosystem. Levolution is the process that, in this general way, creates a new and higher order holosystem on a higher level of organization. This conception has its roots partially in the ideas of Arthur Koestler, who recognized the whole-part duality of the systems in nature.

The Universal Functions of Holosystems

Like all dissipative structures, holosystems are compliant with the principles of thermodynamics. These principles of generalized holosystemic, or entropically functional, order are recognized here as the *universal holosystem functions*. They comprise a description of the very special energy flow pattern of functional order.

a. Holosystems must capture energy from their environment.

b. Holosystems must use energy to maintain their order, do their work, etc.

c. Holosystems must degrade and dissipate energy (as entropy) to their environment.

d. Holosystems must dissipate energy maximally, as fast as possible within applicable constraints, which includes their survival.

e. Holosystems must have boundary conditions and be distinct from their environment.

f. Holosystems must be composed of parts, and those parts must also be Dissipative Structures.

g. Holosystems must allocate or distribute energy to their parts.

h. Holosystems must receive, manage, and coordinate the work, produced forces, or useful energy output contributed by their parts.

i. Holosystems must manage and dissipate the entropy dissipation of their parts.

The formation of a new holosystem is spontaneous, but that is only the thermodynamic meaning of the word "spontaneous" which means downhill in terms of energy potential or entropically favored. The initial formation of a holosystem is actually accomplished by the process of Levolution, covered below, which spontaneously builds a new holosystem from a set of whole holosystems that it essentially evolves into its parts.

The Whole-Part Duality Principle
The Whole-Part Duality Principle states the obvious, but profound, situation of each holosystem in the universe. Each exists as a whole-part duality. Each holosystem is simultaneously a whole that is composed of parts, and a part of a larger whole. There are thus three levels of organization of direct relevance to any holosystem. The level below, the level they are on, and the level above, of which they are a part.

Building holosystems from the holosystems on the level below would logically always result in this pattern. Its significance is that the "positional ecology" of each holosystem is now much more visible. Systems in environments can now be seen as parts within wholes. It is a truth that Darwin could not have seen, due to the fact that the eco-system had not yet even been recognized as a system, and certainly not as a higher order holosystem.

Parts evolve from wholes by natural selection performed by the dissipative energy flow of their wholes. This is no less true of organisms as parts, evolving inside ecological wholes, than it is of smaller or even fundamental particles evolving inside the wholes represented by the larger composite particles. I will assert that it is also essentially true of gravitational parts, when they collide with a larger gravitational system, but this will require some new concepts. Wholes provide the environment, the energy, and the entropy sink for their parts. Parts not only provide the material for their wholes, but their energy outputs actually do the work for their wholes. It is not the way that we are accustomed to thinking, but it is the profound truth of the situation.

More Perspectives on Holosystems
Among dissipative structures or holosystems, the entropic mandate of thermodynamic law imposes a universal set of individual perspectives that parallel with familiar instincts and tendencies of living organisms. Such systems will tend to seek their own survival, for example. It is certainly not that they do so consciously; it is simply that if they don't pursue "behavior" that results in their survival, they will not survive. The seeking or pursuit is not a matter of choice by the holosystem, it is a mandate from dissipative structure requirements. The continued stability of a holosystem, if the holosystem is to remain in existence and transfer energy downward as fast as possible, is simply part of being a holosystem. They don't generally have the latitude to decide not to survive.

Entropic Drive, a name I loosely give to the "push" of energy in its big hurry to get down to equilibrium as soon as possible, is what its driving the universe, but it is not the only mission operating here. Holosystems are dissipative structures that get their mission in the universe directly from the Entropy Law and the Maximum Entropy Production Law.

Entropic Drive just wants to get energy lower in potential, degraded and dissipated. It seems like a much higher purpose to perform order creation, to create dissipative structures and holosystems, but these phenomena exist only because they aid in the universal mission created by the two Laws relating to entropy. The reason that the universe does what it does, is because it is full of holosystems that are on the same mission, but the ultimate reason they exist is due to the Entropic Drive itself, and it is created by the operation of the thermodynamic laws. It is apparently the case, as hard as it is to accept, that it is only a secondary side-effect that the universe is producing an increase in holosystemic, functional order. Such order is like a strategy of the universe's entropic drive.

From the perspective or frame of reference of holosystems themselves, continued survival, not measurements of energy flow, are most important. Thermodynamics is fundamental, but from the perspective of holosystems, survival may be even more fundamental. Efficiency may serve survival better than more power in some cases, and some holosystems, like us, can even think and decide for themselves.

As humans, we tend to adopt and recognize the perspective of the holosystem itself rather than the perspective of the energy in it. Energy always simply takes the steepest way downward, but it may be selecting among pathways that are actually sentient holosystems. The amount of entropy that a dissipative system may produce depends on a number of things. It will certainly depend on its existence as a dissipative structure (i.e. it's survival or not), but also on its size, scale, scope, and range, and these are technically related to its power or work output per unit time. From energy's perspective the bigger the flow the better, but for a holosystem on any given level of organization, the game changes to one based on survival. On some levels the game also includes reproduction, but it is easy to see that this is, at its base, really nothing but a fancy way to both promote the survival of the type, and provide a source of random variation on which natural selection may act.

The following few perspectives are not intended to be understood as thermodynamic laws, but they are the way that the mission of entropy production is translated into parameters of natural selection and how it affects the frame of reference or the perspectives of dissipative structures.

The Survival Perspective: A dissipative system will tend to seek its own **survival** because the alternative (non-existence) will not dissipate energy as rapidly.

The Growth Perspective: A dissipative system will tend to seek an increase in size, scale, scope, speed, and/or range to maximize the amount of energy it can capture, use, and dissipate. Growth can be size growth or population size growth.

The Power and Efficiency Perspective: If energy is relatively plentiful, a dissipative system will more often tend to maximize its power. If energy is scarce, a dissipative system may more often tend to maximize its efficiency, which enhances the survival of more member of a population, or the same number over a longer time.

Power is the amount of useful work done per unit time. Energy does not care if its flow is useful or not, but a holosystem is suddenly a system that has interests that are, or at least may be, different from energy at large (Salthe, 2010). Power and efficiency can both be important for survival, but they cannot be maximized simultaneously.

Facultative maximization of power is highly adaptive, but not all dissipative structures can make facultative choices. Maximizing power sacrifices efficiency, and where energy is scarce, a shift from maximizing power to maximizing efficiency extends the usefulness of the environmental resource, and this can relate directly to system and population survival. In another sense, however, this same choice may operate to shape the properties of evolving holosystems in differing environments, whether or not an individual is capable of a facultative switch. Evolution tends to act as if it is facultative in this regard and makes the choices on behalf of the holosystem if need be. This reality is the source of the Stoic concept, also present in several existing religions, that one may not chose the right path, or even may chose not to follow the right path, but one will then be dragged along that path just the same.

The thermodynamics of the survival perspective of holosystems can influence or affect the operation of The Maximum Power and Efficiency Principle and affect the flow of energy. Energy may mandate maximum entropy production, but the dissipative structure's own order and "instinct" for survival may cause it to interpret this mandate as either surviving scarcity to maximize energy dissipation over a longer time, or risking its long term survival to maximize its useful work more immediately. In this sense, holosystems have a new emergent perspective in survival.

Law of Functional Order #6 - The Law of Holarchic Unity

Through the process of Levolution, new holosystem types have always emerged, and they do so in a continuous sequence of levels of organization, displaying a general increase in functional order over time. The levels of organization, which are connected by nodes in a part-whole relationships ranging from the massless photon to the galaxy and everything in between, represent what I will call, in following Arthur Koestler, a holarchy.

Holarchy is a much better term here than "hierarchy," which many authors have used. Not only does it allow a clear indication of the subject; it also connotes the notion of the wholeness of each holosystem and their nested relationship, while avoiding the misinterpretation of hierarchy as a command and control structure in organization theory, which mainly represents, and certainly connotes, differing authority, the power of control, and the related status.

In a way, the holarchy, as a concept of a structure inclusive of all of the order of the cosmos, is the obvious complement to the concept of the "Monodyne" of energy. The holarchy refers to the edifice of multiple levels of order, where the Monodyne refers to the energy itself which flows through it. The Monodyne is the single batch of energy that arrived as the Singularity, and animates our universe today. It has clearly differentiated into forms even as it has become arranged and ordered into a vast array of dissipative structures and holosystem types, but it was one batch of energy at the Singularity, and it is still one batch today. Every tiny volume of space contains some of it. Energy flows in the patterns of a myriad of holosystems that comprise the unique structure of the holarchy of so-called material structures. Energy has cascaded downward, through a sequence of dissipated and degraded forms, and it has thereby differentiated through degradation. Herein lies the importance of the Law of Holosystem Unity.

The science and the thermodynamics of the holarchy and the universe as a monodyne, if one can get past the philosophical heft of these phenomena, are about the simple fact that these two components of the universe are indeed unified. Levolution constructed the holarchy of material order even as the monodyne of energy has differentiated and degraded. Each level of organization utilizes one or more forms of energy, but each is in the business of degrading energy and each level usually degrades it on a larger scale. In other words, Heraclitus had it right. *All things are one.*

Energy and order, as difficult as these subjects have been to investigate and understand over the centuries, actually have maintained their unity as they created the cosmos and its contents. The holarchy and the Monodyne represent the cosmic duality of Order and Energy. Because Order is a pattern in the flow of Energy, these two things are really one thing; energy. But without understanding energy's relationship to order, we would not see the great apparent "wisdom" of energy in its inexorable downward flow. I am waxing philosophical, but this has to be one of the most profound and insightful views that one can have of this universe. The profundity of its inclusiveness and unity can hardly be overstated.

And so this chapter on the new Laws of Energy and Functional Order recounts my bold proposal for an amendment to the Thermodynamic Laws, to include the proposed Law of Maximum Entropy Production, and to add another whole set delineating the six Laws of Functional Order. These concepts are the heart of the Paradigm of Levolution.

Basic Levolution

The universe as we know it is the result of the growth in entropically functional order. The actual process that builds holosystems and creates functional order is Levolution. It is thermodynamically spontaneous, or downhill, but that does not explain it. Levolution proceeds as a process of the evolution of wholes into parts by Thermodynamic Natural Selection, but here is another new perspective.

The environment to which holosystems adapt is an environment that shares many aspects in common with the evolving holosystems themselves. The environment is a larger, but structurally self-similar whole, which will essentially define the niches and the fitness of its parts. In fact, the so-called environment may already be another discrete holosystem at a higher level.

A correspondence, a coinciding of the energy-moving functions between holosystems on successive levels of organization is what makes Levolution work. This scale-invariant self-similarity is a result of thermodynamic isonomy; equal status under the law.

Energy's entropic drive powers the creation of all dissipative structures, but among holosystems, it takes on a different character. It turns wholes into the parts of larger wholes. Entropy and Dissipative Structure Theory are important aspects of Levolution, but so is the structural evolution of "wholes into parts," and so are some "new" ecological-sounding principles. I will explain the important perspectives that come from general holosystem ecology and Thermodynamic Natural Selection a bit later.

Hints of something like Universal Thermodynamic Evolution and Levolution have existed for a long time. Ancient philosophers have remarked about them. The Great Chain of Being, from the time of Plotinus in ancient Alexandria, was an early and related conception of a stratified order inherent in the universe. With our current, interdisciplinary understanding of science, almost everyone has already recognized that for things to exist in a nested structure on levels of organization, those levels must have formed sequentially. Levolution is the process that did it.

These two theories, Universal Evolution by Thermodynamic Natural Selection on one hand, and Levolution on the other, may be viewed as the merger of thermodynamics and ecology, or the extension of evolution theory to all the systems of the universe. The two processes really do unite our view of physics, ecology, and humanity in a profound way. The change of perspective here affects all of the scientific disciplines, both the natural and social sciences. There is no fundamental separation, either in nature or in our sciences, between nature, humanity, or it's cultures. They all capture and dissipate energy. It is a matter of getting our most basic perspectives correct, and seeing things for what they really are.

Many of the sciences will come into view for us here in a new way, as studies of energy flowing in very similar systems. It is still amazing to me that most of these energy systems obey the same natural laws and share so much in common. Similarities among naturally occurring thermodynamic systems, combined with similarities among their environmental challenges and situations, leads to the broad and universal applicability of the Levolution Paradigm. What started for me as a "mash up" of economics and ecology, has transformed into it.

My objective is to point up the general principles involved in Levolution, and show how they apply to the various system types found in the universe. The result is the outline of a new Paradigm that explains the universe's content as the result of two simple processes; the universal version of evolution and the universal process of Levolution,

The subject is how this reality creates functional order, and it is to be told without reductionism, at least without old-style reductionism. While the Paradigm will reduce a lot of complexity down to simple thermodynamics, it doesn't reduce whole systems to their parts and expect to learn much about the whole. In fact it has become clear to me that that kind of reductionism, which served science well for a time, really only suffers from a lack of the Levolutionary Paradigm. The Paradigm has a perspective based not primarily on the structures in matter, but on the functional patterns in energy flow.

Levolution captures the notion of how very simple the universe is, despite also being vast and complex. It is a new framework based on:

(1) energy and the laws of its flowing which provide the ultimate goal of entropy dissipation
(2) an ideal structural model of a functionally ordered system that promotes this goal
(3) the natural selection process that works to create such systems
(4) the evolutionary process that maintains holosystems over time through adaptive change
(5) the Levolutionary aggregation process that forms new levels of material organization
(6) the resulting multi-level holarchic structure of the universe's matter.

The Levolution Paradigm is the solid core of a new, scientific world-view on which to hang both intuitive common sense and previously acquired scientific knowledge. It is a framework that provides a remarkable view of the evolving universe, as awesome and unified as it really is.

CHAPTER 6

THERMODYNAMIC NATURAL SELECTION

The kingly power is a child's.
The world is like a child playing
draughts (checkers).
        – Heraclitus

THE ASCENT OF NATURAL SELECTION

One of the most central ideas of the Levolution Paradigm is that the process we know best from biology, Darwin's mechanism of natural selection, is not restricted to biology at all. It is actually a universal phenomenon. In Chapter 5, the proposed listing of the most important natural laws related to energy and order, natural selection is the second law of functional order. It can appear there because there is then a pathway selection principle as a law of energy, and only at this point, after dissipative structures, can there be populations of discrete and variable entities to select among.

Natural selection is universal because it emerges from universal and thermodynamic mandates related to energy flowing through populations of discrete holosystems. Given a population of these dissipative structures and energy moving systems, the potential for differential success among its individuals in maintaining their existence, is logically to be expected. It would be very unusual if both the systems and their environments stayed the same. Similarly, variation, either among the holosystems themselves or among the details of their environments, is also to be expected. These are very basic notions, and it seems to me that variation and differential survival may be taken as "givens" of any naturally selective situation. Where these are not found, there simply can be no natural selection.

The challenge of helping physics accept Thermodynamic Natural Selection as a universal law is met by the logic of thermodynamics, as revised, and a few simple observations. The challenge of convincing physicists, cosmologist, and biologists alike, however, requires treatment of a number of caveats and more detailed explanations.

Holosystems are energy-driven systems, tuned to capture and dissipate sufficient energy to pay the cost of their internal order, and when the energy they capture is insufficient, or stops flowing, that is generally the end of the system. Energy storage is an obvious caveat, which I ignore for simplicity's sake. Natural selection is simply the emergent, population-level result of such individual system terminations. It occurs when the differences in survival among individuals are related to variations in the match that must exist between the characteristics of the holosystems and the details of their particular environments.

The Levolution Paradigm's central theme is that it transforms the concept of a subject's environment into the concept of the larger whole, of which the subject holosystem is a part. This implies that the match noted above is likely to be a match between a holosystemic function of the larger whole, and the functional contribution to it made by the subject holosystem as a part.

Whole ecosystems probably function better if their parts, the organisms in them, and certainly the abiotic parts as well, function in moving energy faster. From the most general or thermodynamic perspective, the ultimate causes of natural selection converge on failure to adequately perform one or more the six functions of holosystems.

For example, failure to capture sufficient energy from the environment (starvation) is bound to lead to the holosystem's instability and demise.

Survival of holosystems is linked between levels. When an organism dies, its cells die too, and the biochemical machinery quits functioning as well. There is an obvious downward causation involved in natural selection that supports the idea that the holosystems of the universe are built on a very solid foundation. At any given time, the holosystems on the highest level of organization are keeping all the levels of its parts functioning by its capture and internal distribution of energy.

The converse is also true to a degree, a failing part can bring down the whole edifice of a multi-leveled holosystem. The difference is that natural selection can deal with parts statistically, stochastically, by changing the overall characteristics of the parts through a weeding out the poorer performers. Parts cannot effect changes in the whole that way.

By definition, the causes of differential survival are related to the "fitness" of the variants being subjected to natural selection. The environment of the system, which is the same as the whole of which the subject systems are parts, is usually the agent of natural selection. Dissipative structures and holosystems are products of their environments.

Natural selection is the mechanism through which a change in characteristics occurs in a population of systems. Changes are made in characteristics to maintain a continued match with changes in the environment, which are changes in the containing whole. This is expandable to consideration of new environments, as well as changes in current environments.

As we shall see, the mechanism of natural selection is also expandable to the case of Levolution, where the systems, as independent wholes, change by natural selection into the parts of a higher order holosystem that does not yet exist.

In other words, natural selection is the key mechanism of evolutionary change, as we know, and it is also the mechanism of Levolutionary change. These two processes happen throughout the universe, on virtually every level of organization. In a later chapter, I will offer how photons probably evolved into gluons, for example.

Whether one considers the universe to be alive or not, holosystems are easily recognized as either functioning or not; whether they are capturing, transferring, and dissipating energy. Non-operating, or non-functioning, holosystems are automatically in violation of the entropy law and the maximum entropy production law, and must disintegrate down to the highest level of order that can function and survive.

Why Only Holosystems?

Before diving into a more detailed explanation, let's address the types of entities that are subject to universal Thermodynamic Natural Selection. Why only holosystems and not energy pathways?

The selection principle at work in the Law of Maximum Entropy Production is relevant to, but not quite equivalent to, a Law of Natural Selection. The reason is generally because it addresses thermodynamic selection in flowing pathways of energy. All it lacks is consideration of discrete systems. Natural selection is essentially a quantitative subtraction from a quantified population. Natural selection, in the normal sense at least, does not act on individuals or on continuous flows of energy. To construe natural selection in the way Darwin and others have learned and taught it, it can only apply to populations of discrete entities.

When we add the new Dissipative Structure Law, the First Law of Functional Order, we get the needed discreteness of structures or systems. In some situations, like Be*nard Cells, we even get a population of the fluid convection cells. What we do not get with dissipative structures are easily visible events of natural selection among them. The problem is that the examples of dissipative structures that we know, and that are not also holosystems, are hydrodynamic fluid flow patterns as well as thermodynamic energy flows. While they may be adapting as the Be*nard Cells jiggle in the petri dish, they are not often observed to be subtracted by natural selection.

An experiment where the heat source is only acting on a portion of the total dish environment would potentially let us see natural selection among simple dissipative structures, across a varying environment, but unfortunately, these structures generally conform to the dish. I am reasonably certain that natural selection could act on a population of dissipative structures. The ones I know, however, are certainly not good examples of the natural selection process. While I can offer no observable examples of natural selection among simple dissipative structures that are not holosystems, I will stick my neck out and suggest that it occurred in the distant past.

It is notoriously difficult to demonstrate natural selection, but it has been done many times. The darkening of moths due to the camouflage benefits of smog collecting on tress, or "industrial melanization" comes to mind as a classic example.

Holosystems on the other hand, dissipative structures composed of parts that are also dissipative structures, are more clearly subject to natural selection. Each of the sciences has observed it. We have the whole earth-bound scenario of biological versions of natural selection acting on cells and organisms. Further, R.J.P. Williams (1996) has illuminated many detailed examples of nuclear (elemental) and chemical natural selection.

Gravitational natural selection has been mentioned above. Our awareness of the mechanism is increasing, and natural selection may now be seen among ideas or memes in cultures (Dawkins, 1989). It is even occurring right now as your brain decides which of my ideas to accept. Natural selection is not only universal; it is profound.

Constraints as Natural Selection

In a later chapter, I will discuss how composite particles evolve and levolve by the action natural selection. For now, let me simply say that what physicists and chemists usually call "environmental constraints" are often visible to the initiated eye, as natural selection. Both of these concepts essentially control what can possibly exist in the future, and that is the essence of natural selection. In other words, physicists and chemists have used a concept essentially identical with natural selection for a very long time without realizing it. Natural selection is hidden in their concept of constraints.

The most efficient logic that leads to the Law of Thermodynamic Natural Selection as a law of order-creation is perhaps best explained in terms of the energy and order laws that lead up to it. Given (a) Classical Thermodynamics, (b) the Maximum Entropy Production Law, and (c) the Dissipative Structure Law, the Thermodynamic Natural Selection Law follows easily. With these laws, it is given its energetic selection criterion and its discrete structures to operate upon. Given informally that there may be variation among such structures, and that they will not just occur once, but will exist in populations, the elements of natural selection are all evident.

As I see it, the necessary logical components must appear before it in the list before a given law appears. There is a sequence in the proposed list of new natural laws. The laws of energy and order, more than most, need to be arranged in an order of "fundamentality". In going down the list, each law is about phenomena that are derived from the operation of the laws above.

Thus, we are led to face a fact that is not easily treated. The problem is that natural selection, not to mention evolution and levolution are required to get to the holosystem structure, so it must be simply the dissipative structure model that provides the discreteness, variability, and plurality required in any case of natural selection.

The problem then becomes the observation that it is not tornadoes, swirls, or convection cells that seem to be governed by natural selection; it seems to be the living dissipative structures, the more complex things. This, I will assert, is an illusion. Such simple dissipative structures are subject to natural selection. The difference is that the notion of "environmental constraints" is more typically used instead of natural selection, in these realms of physics and chemistry. Natural selection applies to dissipative structures and does not require its subjects to be holosystems.

Such a revolutionary sounding proposition as Thermodynamic Natural Selection requires that we seek to get it in the right place. We definitely need to finally nail down this important proposition. Natural selection viewed as a thermodynamic law, has been on the table since Alfred Lotka first proposed something like it in 1922. To comprehensively describe natural selection as an extension of thermodynamics, and transform it into a universal law applicable to all holosystems, we must address these subjects which support the proposition:

- Maximizing entropy production by selection of energy pathways
- Dissipative structures as discrete, quantized pathways
- Continued production, durability, and/or reproduction of forms
- Differential termination of individuals in populations
- Sources of variation in a population of dissipative structures
- The generality of thermodynamic fitness
- The part role niches of energetic systems
- The whole as the environment, and the agent of natural selection
- The ultimate source of thermodynamic selection pressure
- The Dissipative Structural Model as the target of natural selection

The above list of topics, caveats, and qualifiers is expanded below, and provides the concepts to link natural selection to thermodynamics, hopefully to the satisfaction of ecologists and physicists alike.

Maximum Entropy Production

This proposed Fourth Law of Thermodynamics provides the actual selection criterion of Thermodynamic Natural Selection. It is based on energy's mandate to get to lower levels of potential *as fast as possible*, within constraints. Recall the analogy of rainwater flowing on a mountainside to understand the general selection mechanism involved here. To dissipate faster, energy always favorably selects and also allocates itself to the pathway(s) that take it downward the fastest and it negatively selects the others. This law applies to energy at large and is not restricted to discrete dissipative structures or holosystems. That is why it is a Law of Energy rather than a Law of Functional Order. Energy itself does not terminate (from the conservation law), and that is why this law of pathway selection needs logical help in getting to the Law of Thermodynamic Natural Selection. It is the next one, the law of dissipative structures, that provides the needed help.

Dissipative Structures as Discrete Entities

The proposed Dissipative Structure Law provides us with discrete energetic entities, which can also be conceived of as discrete energy pathways. Such structures can either demonstrate the quantitative "subtraction" that characterizes most examples of natural selection, or demonstrate the action of constraints in the environment that prevent such an entity from existing. While all dissipative structures are potentially subjects of natural selection, only holosystems, which are dissipative structures, actually exhibit ample evidence of its operation. Tornadoes and swirls at drains do not seem to evolve by this process; they are apparently too simple and too transient. However, the concept of constraints and the concept of natural selection are equivalent in the sense that they essentially *determine* which discrete entities can exist in the future.

Production and Reproduction of Dissipative Structures

One of the most basic principles of dissipative structures is that they form spontaneously under certain conditions of increasing energy flow. This phenomenon is most typically thought of in the example of convection cells, ordered hydrodynamic vortices, that form spontaneously to transfer heat, or dissipate energy, at a faster rate than before they formed. Production of holosystems, because they are dissipative structures, follows this pattern, but before the biologists reading this get too agitated, I must address reproduction.

When autocatalytic reaction sets were evolving on earth, they faced both the opportunity and the challenge represented by the millions of possible chemical combinations of atoms and molecules in the various nooks and crannies on the earth. Marshaling these into intricately sequenced chemical reactions with the property of autocatalysis seems difficult for natural chemistry to accomplish, but there was a lot of time (Kaufmann, 1995). Add a template for assembling the needed proteins, and a way to deliver them in the right sequences and amounts, and it becomes doable. The wheel need not be reinvented every time one is needed.

But further, this template for building an organism, automatically becomes a tool for multiplying that organism more rapidly by simply reproducing the template. The genetic template became the earliest known instance of naturally-encoded information representing entropically functional order. It is information about how to build a holosystem.

Reproductive success, and not mere survival, then becomes the way to fitness, which is the converse of natural selection pressure. Biochemistry is a well-ordered phenomenon indeed, and the major innovation supporting the biological realm, but it is not required to apply the principle of natural selection. Reproductive success became the biological measure of fitness because the biochemical challenge faced by the chemistry of life is substantial, and flowing energy has a capacity for innovation that is enabled by natural selection.

Differential Termination of Structures in Populations

The thermodynamics of system demise among dissipative structures or holosystems must be treated. This phenomenon is what happens when the energy inputs are insufficient, or the required energy functions of the dissipative structures are not fulfilled, or for any reason a dissipative structure or holosystem becomes terminally and thermodynamically unstable. The result is termination and disintegration of the dissipative structure, down one or more levels to the level of parts that can survive. This is individual death, but natural selection is not about individual death.

In a population of such structures however, there may be variations among them. Differential levels of system demise, or differential reproductive success, among these individuals, if it is related or correlated with their variations, results in natural selection. The selection may favor some and be against other variations, and the "characteristic profile" of the population changes as a result. Natural selection is a population level phenomenon, which is to say that it is emergent, like the pressure of a gas is emergent among a population of moving particles.

Sources of Variation in a Population

There are many sources of variation in any population of dissipative structures. Brooks and Wiley (1988) provide a good contribution here for the biological realm. Lyn Margulis did as well, adding "endosymbiosis" as a process that can yield innovation in the form of new composite forms. Endosymbiosis is a good example of Levolution, so chalk it up, if you are counting.

At the root of all this, space and time represent the dimensions of possible holosystem differences in a population. Varying proximity to boundary conditions, resources, wastes, and energy fields create many important differences in the environments of holosystems. Any source of variation in populations of either holosystems or in their environment is sufficient to result in natural selection.

The Concept of Thermodynamic Fitness

Thermodynamic fitness is a very interesting concept that has barely been considered at its depth. Ultimately, any competition in the struggle for existence of any holosystem, whether for resources, energy, or mates, may make anything in the universe into a "bone of contention". The ultimate finality of thermodynamics allows us to see that *virtually any selection criterion is ultimately going to reduce to thermodynamics*.

In any competition for essential resources, losing aligns with scarcity, termination and disintegration, while winning aligns with fulfilling essential needs, maintaining thermodynamic stability, and surviving as a dissipative structure or holosystem. More could be said, but there is probably no need.

The Thermodynamic Part-Role Niche

The logic of thermodynamic fitness may be sharpened through a view of Charles Elton's original, "role-based" version of the ecological niche (Elton, 1933). This is to be contrasted with the more common notion of the Hutchinsonian ecological niche, which is a multidimensional hyperspace composed of regions enclosed on dimensions of variability that may exist in an environment.

The thermodynamic niche for holosystems may be viewed either way, but it is perhaps best viewed as a role-based niche concept, but here we restrict it to thermodynamics, and relate it to the roles that fill the definitional needs of dissipative structures, and constitute the six universal functions of holosystems.

Just as species have known roles in their ecological communities, the thermodynamic roles of holosystems are also known at a general level. They are the universal functions of holosystems treated earlier. This means that thermodynamic fitness, which is always assessed by the environment, or the whole of which a subject system is a part, is related to its function within that larger system.

Fitness in the Levolution Paradigm is always related to the very "reason for being" of the part. The reason for being, from the perspective of a subject part is to be found in the needs of the larger whole holosystem, that our subject fulfills and performs perfectly, because it is well-adapted as a part. Of course if it is not, we already know what will happen.

The Whole or Environment as the Agent of Selection

As noted above, the agent that delivers the service of natural selection is the environment, and the environment is the whole holosystem on the next level up, of which a subject holosystem is a part. There is no mystery here, but there is also no limitation. Because all holosystems generally occur as populations, and their types are arranged in levels, every level of holosystem arguably experiences the phenomenon of natural selection in this way. The differences are in the details.

The Ultimate Source of Thermodynamic Selection Pressure

As I like to repeat for emphasis, there is a primary and ultimate mission of energy flowing in the universe. It is to dissipate as fast as possible. This "entropic drive" which is energy following the regularities we call natural laws, seeks to obliterate energy potentials, and dictates that energy will always do anything it can (within constraints) to dissipate faster. It will even "explore" organizational landscapes at higher levels that make energy degrade and dissipate faster. That is why it forms many levels of holosystems, each larger in scope than the last. Holosystems are ultimately judged based on their ability to transfer energy downward in potential and size definitely matters in this endeavor.

The Holosystem as the Universal Target of Natural Selection

The old innovation of energy flow that is the holosystem pattern, as it has been implemented similarly on many levels of organization, can even be seen as *the general model of thermodynamic perfection* that natural selection works to carve out. In its generality, the holosystem pattern is usually the ideal, the desired form. A dissipative structure composed of dissipative structures is an ideal situation. If your environment changes, you change with it. If your parts need to change to effect the adaptation, they too can change. I can see why nature shows a preference for the universal holosystem structure. It is a matter of thermodynamic law, and it is inherently very functional, given the entropic drive for order creation and maintenance.

Given energy flow, the adaptive process of natural change is always trying to create larger and larger holosystems out of whatever it may have to work with. It is usually the case that what it has to work with are existing holosystems, and so we see the phenomenon of Levolution.

So the logic that leads to Thermodynamic Natural Selection as the Third Law of Order is completely contained in the above logic. The major supports are the new Fourth Law of Thermodynamics, or the Maximum Entropy Production Law, and the dissipative structure Law. Once discrete structures exist, holosystems can exist, and the latter might remain stable over long enough periods to encounter the struggle for existence and the reality of natural selection.

THE TYPES OF NATURAL SELECTION

A Union of Disciplines

Evolutionary ecology provides important supporting concepts in support of natural selection as a thermodynamic law. Alfred Lotka in 1922 only provided hints and supporting statements, but did not provide a finished theory. Howard Odum in the 1960's, supported the concept as well, and he illuminated functional ecosystems as energetic, thermodynamic systems.

The key to understanding why ecology is related to physics is to recognize that once "systems" are allowed to be produced by the thermodynamics of dissipative structures, a sort of ecology is immediately applicable regardless of the subjects. All systems have environments as part of their definition, and the relationship between these is the basis of the science of ecology. The union in logic is not so surprising when that is understood, but we can see that a conception of universal Thermodynamic Ecology is needed, just as a concept of Universal Evolution was needed.

While thermodynamics has been around since William Thompson (Lord Kelvin), just prior to the time of Charles Darwin's publishing *On the Origin of Species*, the understanding of energy flow within dissipative systems was slow to follow. As Ilya Prigogine himself pointed out, it is the classical emphasis on time symmetry, on reversible reactions almost exclusively, that blinded classical physics to the thermodynamics of irreversible actions, of entropy, dissipative structures, and increasing order among systems far from equilibrium. That is particularly ironic because truly reversible reactions do not really even exist.

Holosystems are Real

In this book, I have introduced the construct of the "holosystem" as a special kind of dissipative structure that is composed of dissipative structures. I now would like to switch gears somewhat and support the idea that the construct is really much more than a theoretical construct. It is a large category of real entities, and may be the largest category of things in the universe. The construct is real and holosystems are real. They have populated the whole universe.

The holosystem pattern is the teleological goal of natural selection as it performs Levolution. The reason this is so is because the acceleration of energy flow downward in potential is the entropic teleological goal of the universe. As much as I hate to say it, holosystems represent the implementation of a nested dissipative structure "design" achieved through the process of natural selection acting upon a population of holosystems. Now if your ears pricked up at the mention of a universal design, do not get too excited. The design has no designer, unless it is energy itself, as it seems to be.

The production of novel holosystems is the capture of a flow of energy from a population, and its guidance through the functions of a new, larger scale, dissipative structure. The new design is a function of Levolution, which builds new holosystems, but it is also just the energy flowing in accord with natural laws. Evolution operates to maintain the holosystems in an adapted state. Natural selection, which in a way is simply a phenomenon related to the realities of how they fare, whether they exist, and whether they continue to speed energy along, operates to effect change in a population.

In some cases, holosystems must do their energy-moving tasks better than competitors. In other cases competitors have been left in the dust, and there may be a slower pace to life. Either way, any perceived fickleness of natural selection should have been left in the dust as well, and we can see that it pursues a beautiful model as it does its ugly work.

Ludwig von Bertalanffy would be happy with this concept. His General System Theory (GST) has always been a homeless, useful, but hard to place type of a theoretical construct. If the powers that be ascent to making the generalized holosystem model into a matter of natural law, it will have evolved directly from a GST beginning.

We have, in a way, gone full circle. Energy flowing in populations of generalized holosystems subject to natural selection provides a complete thermodynamic mechanism for speeding along the energy flowing in the universe by building organized systems, and keeping these holosystems adapted to their changing environments.

The nature of it will predictably be a tough pill for physics to swallow. It has been rejected before, and I have read the works of many authors who wriggle and squirm to find ways to reduce the impact of natural selection, or join it with other concepts in addition, or make up fantasies like self-organization, to gloss over the simple fact of it.

One of my main objectives with this book is to get the arguments about this issue correct, and drive this point home with all the communicative power I can muster. Natural selection is a thermodynamic law promoting systemic, functional order, and is the basis for both Levolution and evolution, which are also essentially thermodynamic laws.

You should make a judgment yourself, as the question is probably a new one. If you ignore the question you will not reap the benefits of the answer. Natural selection, as a thermodynamic universal, is the truth that sets the universe free. Even if you do not see the universe as "alive" you should prepare yourself to see that natural selection is the primary mechanism behind its functional order creation and its further evolution.

Natural Selection and its Mechanisms

Dissipative structures and holosystems become unstable when they do not capture, use, and dissipate energy in such a way as to maintain their pattern in the overall energy flow. At that point of instability, they cease to exist, or terminate. Each such system must be viewed in its own frame of reference, in its own whole or environment, and with its own energy resources. The holosystems will vary and environments will vary. Some systems will remain stable, while others will not. System termination, and similarly survival, is thus differential within a population. This causes change.

Differential survival is one of the ways in which the philosophically important dichotomy between existent and non-existent is achieved. Three other ways of achieving this dichotomy will emerge in our discussions.

One is that existence may be prohibited and disallowed by natural law, which creates constraints on what can possibly exist in a given environment. Another is that, in the dimensions of space and time, an existing system may simply disappear from a frame of reference due to its motion. Frames of reference become especially relevant when it comes to the gravitational form of natural selection. The third way is the one that Heraclitus and Plato both dealt with, the phenomenon of a change in form. When a holosystem changes form, the initial form ceases to exist, while the new form begins to exist. We do, and do not, wade into the same river.

This latter way of expressing the dichotomy between existent and non-existent, a change of form, would be troubling for us if it were not for one thing; change of form is the subject here. The phenomenon of evolution by natural selection does not need to happen, and probably will not happen if a subject system can inherently adapt by changing its own form. The ability of a holosystem to change form is the ability to avoid natural selection.

We may thus eliminate this kind of *direct change of form* as a relevant mechanism related to natural selection. We should not lose sight of the fact that while a whole system may be able to change form, there are likely some lower level aspects, or parts of the system, that are selected in order to effect that change of the whole. Form change of a single whole system may be accomplished by the natural selection of parts. Indeed this is exactly the revelation required to understand Levolution, but that is not our subject here.

So we want to look at natural selection very fundamentally, and understand that it is a universal phenomenon of holosystems on all levels and it is related to their characteristics and their future existence as a fit part of a given whole.

THE VERSIONS OF NATURAL SELECTION

The Particulate Version – Survival of the Stable

Natural selection as it applies to sub-atomic particles, which exist in fields of energy of various forms, is a simple matter of maintaining thermodynamic stability within the constraints of their environment. I do not have any special knowledge here, but for particles, existence and thermodynamic stability are virtually synonymous. Particles become unstable when the forces or fields that hold them together break down.

Particles are aggregated into composites by attractive forces. Anything that turns off or reduces the forces can result in natural selection among varying particles. There has been plenty of time for all this to settle out, and a detailed speculation about the early particles will come in the next installment of the series. The challenge, it seems, is not just to understand how particles might be selected against, and go away, but how they would maintain their population sizes in view of a continual subtraction by natural selection.

The answer is not multiplication, as it is in biology, but continued addition, and the fact that stable particles in stable environments may last a very long time. In other words, the conditions and constraints of continued creation either exist and replacement particles form, or they don't, so they don't. If they do, the disintegrated particles may be subject to the same forces and reassemble, which is to say they may be replaced through multiple, repeated events of Levolution, which is the likely mechanism of any holosystem's initial creation outside of a reproducing situation.

If the conditions of creation do not exist, the population size will predictably be reduced by natural selection. It is also worthwhile to note that the conditions and constraints of continued creation could themselves change, and this would result in variation upon which natural selection can then operate.

Speculation like this can never prove anything, but natural selection among sub-atomic particle types is considered very likely. Any evidence that particles have changed characteristics over time would be evidence in support of natural selection. Dissipative structure formation is the exclusive source of functionally ordered systems in the universe, and natural selection can be seen to be the source of their change in the universe. Any change observed would at least be weak evidence of natural selection.

The Gravitational Version – Survival in Orbit

The only version of natural selection that I really had to figure out on my own was the gravitational version. When I finally saw the process, more than a few insights rushed into my mind. Gravitational systems are subject to natural selection, just like everything else.

Gravitational bodies or systems have a localized frame of reference, outside of which anything that happens is irrelevant to them. When one gravitational object (A) is just flying by in space, with a velocity and mass (momentum) sufficient to allow it to escape the gravitational attraction of another body (B), it will simply exit the frame of reference of body B. From the frame of reference of body B, body A may be said to have been naturally selected against. Its demise as a relevant holosystem or as a member of the population of local gravitational holosystems is not a normal demise, but object A is gone nonetheless, and totally out of the picture. From the perspective of B, this fate of A is the termination of its relevance, which can be seen as a natural selection based on its characteristics of position, momentum, and direction being simply unfit to achieve orbit around B.

The other two obvious options based on gravity, position, and momentum, are that either A goes into orbit around B, or A collides with B and merges with it. In the case of an orbiting A around B, A is a survivor. It maintains its independent existence, at least temporarily, because its characteristics match the fitness parameters dictated by the gravitational attraction of B.

In the case of a collision and merger of A with a larger B, A is selected against, and B is a larger gravitational composite. Body A no longer exists as an independent gravitational holosystem. The mass contribution of A, when it merges with B has consequences, but these consequences do not change the fact that A has been selected out of independent existence as a gravitational body, or the fact that the future is now all about the new B.

Gravitational natural selection is this simple, with selection criteria from Newtonian physics, and it represents the best hope of reducing natural selection to mathematics. The criteria are simple, quantitative, and predictable.

While it may be hard to accept that this simple scenario is on par with the profound phenomenon of natural selection, which causes evolution, and Levolution, this is nonetheless the case. Gravitational bodies survive or not based on a multitude of scenes like this, and that is all it has taken to build most of the structure of the cosmos.

Gravitational natural selection has strange fitness criteria by comparison with most other realms, but Universal Thermodynamic Evolution is demonstrated by it, as is Levolution.

When bodies gravitationally aggregate into larger composites, a very primitive form of Levolution is occurring. Masses are aggregating, and working for a new center of gravity, for the purpose of making more energy flow downward in potential. When physicists talk of anti-entropic properties related to gravity, they are talking about the fact that mass is coming together and not dissipating.

The growth of the gravitational form of functional order is sometimes called anti-entropic. This is because it is obviously aggregating matter into a less random, more ordered, arrangement. We can now see, however, that this material aggregation is really very pro-entropic. The gravitational holosystem is a dissipative structure. Gravitational holosystems operate in accord with the same laws of functional order as everything else. Up until now, however, we have not seen the energy capture, or the entropic dissipation of gravitational objects.

As we will learn in the next book, which is a speculative, but consistent, and revolutionary cosmology based on Levolution, gravitation is the cause, and not the result of mass. In that installment of the series, we will learn that gravitational holosystems consume the energy of space, their work is gravity production, and their entropic dissipation of energy is degraded space. Luckily, the degraded space is itself consumed within normal matter.

The Chemical Version – Survival within Constraints

Atoms and molecules and the reaction vessels, natural or otherwise, which are their environments, and the families of reactions which may interact and catalyze each other in some situations, are our subjects here. Chemical natural selection is quantitative and is well known to be based on thermodynamic stability, reaction potentials and rates under various energy regimes. Products and reactants come to steady states and varying proportions of them are the result. Some possible molecules in these common events have been selected against. Others have been favored. The environment has decided the issues, but the issues may be chemical, as in which kinds of atoms are present and in what amounts, or may also be thermal or kinetic.

Natural selection in chemistry can be accomplished by modifying the parameters of temperature and pressure in the various milieus that naturally vary around the universe. These parameters are involved in determining the phase of matter (solid, liquid, or gas) dictated for a given molecule by its reaction environment. Solids may not combine easily even though the atoms would combine if dissolved in solution. Density differences may keep liquids from mixing, and it is generally the case that chemical mixing is a prerequisite to chemical reactions. Most of these considerations are matters of simple thermal-kinetic order.

Earlier reactions may have already produced some molecules and taken the reaction products to a plateau of stability from which they will not budge. A reaction has been prevented, hypothetically. Sometimes energy must be added to get a reaction to occur. Lack of energy in that case would prevent the products from forming. Things like the temperature, pressure, or the availability of energy are more typically called environmental constraints. Like natural selection, environmental constraints govern what can and will exist in the future. As noted, constraints are virtually synonymous with natural selection, but not exactly, because their operation does not require a prior existence of the form selected against.

These are the common forms of chemical natural selection. While chemists do not generally use this mindset, they do not ignore the fact of it. The interested reader is referred to the book *The Natural Selection of the Chemical Elements* by Williams and Da Silva (1996) and to the works of Stuart Kauffman.

Biological Version – Survival of the Fittest

Darwin's natural selection is alive and well. The innovations of reproduction, fixed duration lifetimes, and generational patterns in biology are a suite of novel mechanisms that only pertain to biology. These lead to the criterion of reproductive success, as opposed to mere survival, as a better measure of biological fitness. It is noted, however, that survival to reproductive age is still a part of the biological version, as is the survival of the offspring.

While the biological version has advantages when it comes to rapid colonization of new territories, recovery from population crashes, and magnification of the effects of an evolutionary improvement that is minor in its scope, it is a recognizable enhancement, and not a violation of standard, survival-based natural selection, as it often occurs on other levels of organization.

While it is proper for biologists to note the differences, it is proper here to note their similarities. Thermodynamics is involved in biological success, no matter how one slices it. Organisms are all about the flow of energy in ecological roles. Even though the criteria involved in natural selection may range over many things, each of those things can ultimately be followed to how they either do or do not help an organism move energy through itself and its environment.

We should not ignore sex and sexual selection. The biological innovations noted above relate to reproduction. Sexual reproduction provides for greater opportunities for recombination of genetic alleles (pairs of genes which produce different phenotypes) in novel patterns. In other words, sex is an additional source of innovation and variation. Sexual selection typically turns females into the agents of natural selection for the characteristics of males. Females control reproduction in this way across much of the animal kingdom, and as Darwin noted, this leads to some strange looking males. Peacocks come to mind. Don't worry however, underneath the feathers is still a holosystem consuming energy.

Natural selection, as biology has found, is a population level phenomenon, an emergent property that is equivalent to the differential demise of variant individual organisms. In biology, death prior to reproduction strongly disfavors genetic representation of that type in the future. The ability to reproduce is a biological innovation, and outside of biology, at least toward the lower levels, reproduction is not really an option, and is not central to natural selection's operation.

While all holosystems may be temporary, in biology the generational system, and more or less programmed lifetime durations, combined with reproduction, all allow for an improved mechanism of evolution. It is still Thermodynamic Natural Selection, but it is set up for maximal effectiveness when it is augmented by the criterion of reproductive success.

Outside of biology, in particle evolution, elemental evolution, molecular evolution, and gravitational evolution, the fundamental survival of the types is still important, but reproductive success is not generally the relevant mechanism.

The cusp between biology and the social sciences is characterized by populations of organisms that organize into societies, like ants for example, and by humans and their cultural organizations which are today organized by the ideas originally created in idea-producing, human, brains. Both of these are instances of Levolution. Ant societies and human cultures are both higher levels of organization formed by natural selection acting to produce larger scale holosystems, as colonies represent.

Cultures are organizations that now feature the production and reproduction of ideas in various brains, and in various cultural institutions, notably education. Ideas are thus the second instance of encoded holosystemic order that is subject to natural selection and it is commonly reproduced to allow rapid evolutionary change. We see cultures rise and fall, however, and what we are seeing in history is natural selection based primarily on the fitness of the ideas that have guided the historical cultures. Ideas about how to live together have replaced the genetic relatedness, which was the old way. Their selection is so much less painful.

The Cultural Version – Survival of True and Functional Ideas

Those who have followed Richard Dawkins know about memes. Memes are basically ideas and behaviors, and while Dawkins makes further distinctions about memes, I chose not to. To me it seems that all ideas are subject to natural selection. When we adopt an idea or believe it to be true, we make it a part of our belief system. When we identify an idea but do not believe it, we are performing natural selection on that idea, at least in our own mind.

In such case the idea does not easily become a part of the whole represented by your belief system. In other words, belief systems are the environmental wholes of believed ideas. Ideological fitness is a perfectly understandable idea. Ideas that are not believed by anyone are essentially dead, but they may not be totally dead if they are encoded in a library somewhere. Consider, for example, Mendel's ideas about inheritance, which were long forgotten.

Human cultures are products of living together socially and the ideas that humans living together produce. Constitutions, laws, regulations, mores, norms of behavior, rituals, belief systems, etc. are some of the categories into which we place ideas. Ideas are selected in the minds of individuals, but this emerges as consensus in a population of minds. This is basically what culture is about. Cultures have several institutions devoted specifically to the natural selection of ideas. Science is one of them. Politics is one of them. Religions are another. Ideas may be innovations and inventions that fill human needs and make people wealthy. Successful ideas are reproduced and broadcast. Ideas that are selected against are ignored.

While guided by ideas, cultures are quite physical in nature. Our culture is really an ecological community, led by human ideas, but containing many species in ecological and economic association. It is an association of an entire trophic web of many kinds of organisms. Cultures whose guiding ideas serve to perpetuate the culture are survivors, but cultures whose guiding ideas are lost, maladaptive, or in serious factual error are likely going to be outcompeted by others. Cultural natural selection may be caused by failure to reproduce the ideas that work in the next generation. It may also be caused by war or conquest. Cultures have come and gone, and it is instructive to observe how they have both flourished and terminated.

Conclusions About Natural Selection

For our purpose of understanding Levolution from a universal perspective, we must conclude there is <u>always</u> a form or type of natural selection that may be brought to bear upon unfit holosystem types. I subsume all four methods, including the operation of environmental constraints, and the exclusion from a frame of reference, all within the scope and meaning of natural selection. Natural selection, in a purely thermodynamic sense, is the differential termination of varying individual members of a set of holosystems, and the consequent reduction of offending system characteristics, among the parts of that whole.

Natural selection is powerful. It is like the law-giver, judge, jury, and executioner all rolled up into one event; the termination of a holosystem characteristic from a population for thermodynamic cause. Thermodynamic Natural Selection, in all its forms, exerts a direct physical control over the existence of holosystem types. It controls their initial existence, their continued existence, and their existence within a particular frame of reference.

The control exerted by natural selection is not necessarily contained in the interactions between a system and its environment. The action of constraints demonstrates that environmental control of existence can exist, even when the excluded holosystem has never existed. The control is obviously not within the systems themselves for the same reason.

A perfectly good frog will not survive in a pot of boiling water. The control of holosystem survival is squarely located within the environment. The environment is always the main source of the constraints inhibiting existence, and the source of the deadly effects of natural selection. Importantly and finally, the environment is almost always the interior of another whole, as I always emphasize.

In every case, and I say at every level, natural selection is an emergent mechanism based on individual system variation from any source, and the demise of the less fit holosystems. The universal process of Thermodynamic Natural Selection, and the changes that it causes, which we usually call evolution, keeps all holosystems adapted as functional parts for their environmental wholes. Natural selection is also squarely behind Levolution, which only rarely creates emergent new thermodynamic wholes, new holosystems, out of populations of holosystemic parts.

I hope to have convinced you by now that natural selection operates universally because energy maximizes entropy production and selection for the discrete holosystems that move energy faster accomplishes that. It is a non-mysterious, emergent, population level effect and truly a thermodynamic principle related to functional order. The principle can apply to any population of discrete things, to dissipative structures fundamentally, but in actuality it seems to apply primarily to holosystems.

Our culture has been in awe of this simple process from our start. For 2500 years at least we have seen acts of natural selection as the principle power that is characteristic of both kings and religious deities. Natural selection carries profound truths, and because my theme is unification, we will want to recognize that great wisdom surrounds the phenomenon of natural selection.

I believe that natural selection was what Heraclitus called the "kingly power", which is the power of execution, of life and death, and this is similar to taking an opponent's checker piece in the game. This power to select is child's play, but it is truly the primary mechanism behind the creative order-building processes of the universe.

CHAPTER 7

UNIVERSAL THERMODYNAMIC EVOLUTION

*Though my reasoning is com-
mon, the many live as if they had
a wisdom of their own.*

– Heraclitus

EVOLUTION WRIT LARGE

We now have a solid platform for understanding evolution as a universal and thermodynamic phenomenon because we have a thermodynamic understanding of its central mechanism, *thermodynamic natural selection*.

Thermodynamic natural selection is the universal mechanism of evolutionary change among holosystems. The function of this process, *universal thermodynamic evolution,* is to maintain a well-adapted state between holosystems and their environments, between holosystems as parts, and the wholes of which they are a part.

While Darwin and subsequent biologists up to now may have missed that last point, it is centrally important to understanding evolution as it really is. Evolution, in the universal sense, is about the process that makes holosystems change and adapt to a changing, or a newly discovered, environment. The current addition is simply the fact that a holosystem's environment is really the inside of another holosystem. This fact does not disrupt, but it changes our perspectives on the great theory of evolution that we already know. We have collectively pondered it in biology for over 150 years.

This chapter is not about building new levels of organization; it is about maintaining the existing holosystems on a given level of organization. The challenge here is to look at how evolution operates in the many realms that it does, and to put biological evolution in these contexts. Evolution by means of natural selection is the way that every holosystem type changes over time. Change among holosystems is made necessary by the mandate of entropic drive. If the mission of energy dissipation is going to have the benefit of dissipative structures in changing or new situations, change is required.

Evolution and Levolution both use natural selection as their central mechanism of causing change, but these two processes diverge in their outcome. While evolution is the continued adaptation of existing holosystem types to remain as the well-adapted parts of some whole, Levolution differs in being the initial adaptation of a population of existing wholes into the functional parts of one complementary new type of holosystem, on a new level of organization. The function in the functionalism is the process of transferring energy downward in potential as fast as possible.

Evolution is Adaptive Change

Universal thermodynamic evolution is a complete theory of how adaptive change occurs among all holosystems. There is almost nothing that evolution cannot operate on. It operates on gravitational structures, sub-atomic particles, atoms, in chemical reactions, in families of chemical reactions, in prokaryotic cells, eucaryotic cells, multi-cellular organisms, societies, cultures, ecological communities, and at least one planetary ecosystem.

My argument for Universal Thermodynamic Evolution (UTE) is mainly supported by the fact that this process of adaptive change, is already a known result of Thermodynamic Natural Selection (TNS), which is the Second Law of Functional Order. Evolution, in the universal, thermodynamic sense becomes the Fourth Law of Functional Order. If TNS is the mechanism, UTE is the result; adaptive change in a population of discrete entities.

Adaptive change is necessary in very basic situations encountered by all kinds of dissipative structures and holosystems. The most obvious type of situation calling for evolution is when the energy resources that have been relied on in the past, are no longer available. The holosystems that came to exist because they could feed on the diminishing energy resource need to change if holosystems are to exist in the new situation. It was food choices that led to the differentiation of Darwin's finches, it is energy choices that are forcing human cultures to evolve.

It would be very mistaken to leave it at this. As was discussed above, any aspect of a holosystem's environment may become a bone of contention, and any of these can be reduced down to thermodynamics. When change is at hand, evolutionary, adaptive change is often nature's answer.

Particle Evolution

While I can produce no direct physical evidence that sub-atomic particles have evolved and adapted, there is some logical evidence. There is only one explanation for functional order, so there is only one way to create particles in the first place, and that is dissipative structure formation. Dissipative structure, or holosystem, formation is the only thermodynamically legal way to produce the kind of functional order exhibited by particles.

Their adaptive change over time or evolution would require only that there was variation among them, that they occurred in populations, and that the varieties met with differing success in maintaining stability as a particle. Here I should note that the common criterion of stability in particles is thermodynamic stability, and we can easily see that this is particulate natural selection in operation.

The most easily considered difference existing with particles seems to be that they have lifetime durations that are very short outside of a given composite particle. The higher order composites serve as the environments of some particles. They are stable inside them, but may last for only a small fraction of a second on the outside. This can easily be interpreted as environmentally induced natural selection, but it is admittedly difficult to call it evolution.

We will have to know more about the particles, and discover acts of evolution among them, to prove that it has happened, but in the upcoming cosmology of the next book I will make a speculative case that it occurred to evolve gluons out of photons.

Gravitational Evolution

Gravitational holosystems abound and are visible everywhere you look on a clear night. Every independent gravitational body is a gravitational holosystem. Gravitational bodies evolve by natural selection, as described in the last chapter, and they move in accord with Einstein's theory of relativity. But a reasonable approximation of their characteristic forces and motions have been known for a long time as Isaac Newton's law of universal gravitation, which describes the force of gravity quantitatively, as a function of masses and the distances between objects, and introduces the gravitational constant.

Gravitational evolution is easiest to understand as the adaptive changes among nearby gravitational objects over time, and one of the easiest examples to discuss is our own solar system. Either gravitational objects find stable orbits in relation to other nearby objects, or the lack of such stability leads to their demise as independent objects in the sun's frame of reference. This gravitational mode of Thermodynamic Natural Selection provides a kind of corroboration that gravitational entities are holosystems. The necessity of relating it to frames of reference is encouraging, because it indicates that the theory is consistent with special relativity.

Gravitational evolution, as the adaptive changes among bodies orbiting the sun, is what is responsible for the heavier elements being concentrated in the rocky planets that are closer to the sun, while the less dense, and gaseous materials exist in planets that orbit farther out. These are two different regions, where a gravitational version of Thermodynamic Natural Selection once acted very straightforwardly to sort matter by density.

The planets evolved and adapted gravitationally into what they are, based on the planetesimals, their parts, in their gravitational environment. In this case, that environment means their radius from the sun in the early accretion disk. The greater attraction of a given volume with greater mass will pull it, move it, closer to the sun. This process of natural selection, defining orbital stability and independent planetary survival based on density, is operated by Newton's laws of motion and gravity, to select different densities of objects for continued orbital survival at different radii.

There is nothing "earth-shaking" about this revelation. Gravitational evolution is simply an important perspective, a new way of looking at the forces and motions of gravitational bodies through the lens of thermodynamics. It does not change the equations of Newton, Einstein, or anybody else. Indeed even if we had suspected gravitational *evolution* before Newton lived, we would have still needed to discover the details.

This, I think, is instructive as a general principle. There are situations in which the Levolutionary Paradigm in general, and Universal Thermodynamic Evolution in particular, are revolutionary, but there are others where it is simply a new perspective. Even where it is simply a new perspective, Levolution offers a single and fundamental unification of the physical principles that gives us a full view of nature.

Elemental Evolution

There is a whole periodic table of the elements whose arrangement was initially organized by Dmitri Mendeleev. They formed by the joining of protons and neutrons into deuterium, helium, and into nuclei of various sizes, and the number of protons gives us the atomic numbers up to about ninety-two. Their place of origin is known to have been, for lighter elements, in the crushing gravity of stars. As the nuclei get bigger, their environment of origin must get even more energetic.

Heavy elements, like gold and uranium, can only form in a supernova explosion.

In other words, the pressures necessary to force the protons and neutrons together into larger nuclei must overcome a repulsive force acting on these nucleons, and it must also trigger or impart something that allows them to stay together after the star blows everything out into a dissipated gaseous mixture. It is likely, I think, that the boundary condition and force that these high pressure regions initially overcome is the repulsive weak nuclear force, and once in range, the reason they stay together after the explosion is due to the strongly attractive, but short-range, strong force.

In other words, the diversity of the elements, the types of stable atoms is due to their evolution in differing pressure environments. These various nuclei represent the particles that find thermodynamic stability in various gravitational pressures. In any given place within a star, differing nuclei will be stable. Heavier ones make their way to the core, lighter ones float outward. It's elementary. The gravitational form of Thermodynamic Natural Selection has evolved the elements.

Elemental evolution is the process of nucleus formation and the wholes of which they are a part are often stars. However, the pressure that is either due to gravity or due to shockwaves determines which nuclei form and which nuclei survive.

In still other words, pressure in the environments of nuclei is the agent of natural selection that has resulted in the periodic table of the elements. This is primarily a matter of perspective, and does not change what it is already known. However, there is a physicist out there who needs to get this perspective, because it will contribute to a new idea he or she is working on. Seeing something both new and generally true, even if it is only a perspective, can be fruitful.

Chemical Evolution

I have already noted that chemistry, the many permutations on the combinations of the myriad of elements and molecules that exist, and the various environments in which they exist, is a clear matter of how and where they achieve thermodynamic stability.

The work in support this section has again, already been done to a large extent. Chemistry is known much better than I could ever tell you. For an evolutionary picture of chemistry, however, consider this. The reaction vessels of nature are not beakers and flasks. Nature has to make do with empty space, droplets, puddles, oceans, and the solid matrices of various rocks until it comes up with phospholipid membranes. Temperatures and pressures are not controlled, and early on, only gravitational evolution has imposed much order on the raw materials with which to work. So not only are there a myriad of possible combinations, there are many types of reaction vessels.

Reaction vessels are the environments of chemical reactions. Environments, really the whole contained soup of which the reagents are parts, determine by Thermodynamic Natural Selection which products will emerge. It is here that reaction potentials, gradients of chemical energy, play out and the reactants turn into reaction products in some proportion.

It may be worth noting that the concept that held up physical chemistry prior to Prigogine, was equilibrium. While equilibrium is an important concept, it is not so with regard to the thermodynamics of dissipative structures and holosystems. Equilibrium should be viewed for what it is. It is a way of saying what state some reaction vessel will end up in should it find itself in what may be a rare situation of having zero energy arrive from outside. In other words, chemical equilibrium in isolated systems seems like a temporary and generally artificial state, much like death in fact, in which the energy has been prevented from flowing. This cosmological view of energy flowing is very different.

Our subjects are all dissipative structures existing far from equilibrium, and I cannot help thinking that because all holosystems are far from equilibrium, that state cannot be too important in the grand scheme of things, except as a final endpoint. I will be scolded about this one, and I will probably plead ignorance. I do more or less understand that the equilibrium state is important in the understanding and quantification of even non-equilibrium phenomena.

Here's another more important consideration that seems relevant to chemical evolution. The elements of life on earth are the common elements, the ones formed in typical stars. Only about 22 of the elements are involved in what we know as biological life. Chemical, not elemental, evolution has resulted in the very large families of very diverse chemical reactions that cooperate in each other's catalysis to create the cyclic metabolic pathways, the membranes, and all the chemical machinery of life (Kauffman 1995).

For chemical reactions to come together in this way is nothing short of amazing. But it is also clearly what would be the expected result of energy constantly flowing, of the chemical version of Thermodynamic Natural Selection, and the resulting evolution of reactions, and reaction vessels. Call it an array of environmental constraints operating if you wish, but keep in mind that constraints are really guiding the natural selection of the most energetically favorable reactions, the ones which degrade the most potential chemical energy, but also perpetuate the whole reaction system's survival.

I think this last point is a very important contribution of the Levolution Paradigm. Once discrete, dissipative, chemical structures have formed, keeping them in place and operating on the tons of high-potential molecules there are to degrade, is a matter of survival for the chemical reaction families, or chemical holosystems. Not only must they function, they must function again and again over time. The chemistry that became genetics is largely about time; the time saved by not having to reinvent each chemical reaction system, even when they are destined to disintegrate.

Biological and Ecological Evolution

This is where it all began. Darwin, and Wallace, pieced together the logic needed to account for the diversity of species and their adaptive transformations to new environments by means of natural selection. Darwin's book was the beginning of biology as a science and transformed naturalists into ecologists. Some time later, Gregor Mendel's work on the heritability of traits was re-discovered, leading to the discovery of genetics, and the Neo-Darwinian Synthesis was born. This theory is central to all of biology today.

Since then, the sciences of ecology and population biology have matured and the underlying complexity involved in evolution as it actually works has been generally revealed. In addition, paleontologists and geneticists alike have worked out the evolutionary phylogeny, the tree of life, that depicts the ancestry of many of the types of organisms on earth.

All the science does not detract from the wonder of it. Life on earth has crept into every nook, every cranny, just about every extreme environment we have looked in. Life has changed over time as well, and with the constant pressure of natural selection to improve the flow of energy, despite the struggle for existence and a continual competition from every direction, it has differentiated into the many elaborate and sometimes beautiful forms we observe.

Each of the species, the forms taken by biological holosystems, is functional in the sense of being useful in the process of energy's universal project to dissipate and degrade as fast as possible. They complement each other in ecological communities in ways that are obviously functional in the process of taking the energy of sunlight, and the energy of our planet's core, and organizing into pathways that eventually lead to its re-radiation into space.

The pathways of energy are generally known as trophic levels or trophic networks among species. Each of the species on earth has a part role in their ecological whole; their ecological community. The communities differ as the surface of the earth differs from region to region. Each community has evolved to operate in the context of whatever biotic and abiotic conditions exist there.

The situation can be emphasized in various ways, and many variations on the central theme are available for inspection and understanding by ecologists. The biotic conditions in which most species find themselves is shorthand for saying that the environment of any one species is largely composed of other species. Relationships of predator-prey interaction, of competition for food and other resources, and the relationships of symbiosis, commensalism, cooperation, exist. These relationships are not static; they adapt to changes when they occur, like climate changes and the replacement of one species by another. Even as one species evolves, others may need to adapt to that reality. They either do or don't.

What does not change is the function of ecological communities. The transformations of biological energy are universally in evidence. Everything from the photons to the duckbill platypus is playing the same game of energy capture, use, and dissipation.

Cultural Evolution

The challenge here is describing the human condition without making it sound like we are so very different from other species, yet also capturing the two most revolutionary innovations in holosystemic order encoding that evolution has yet produced. We not only share the planet with the other species; we are parts with them, of the ecological systems noted above. We have earned a certain position in the trophic networks of many eco-communities, and that position is that of apex predator.

A complicating factor is that our cultures are not accurately described as "human" cultures at all. Our cultures are actually assemblages of many plants and animals, many microbes, many bacteria and viruses. Domestication of species has changed most of them, which is to say that they have been domesticated, or adapted by evolution, into a life dominated by humans. What we call human cultures are truly ecological communities.

We are omnivores, and there are a lot of us. We help energy get to lower levels of potential, and we are playing the same game as everything else in that respect. In fact we have tapped into pools of energy that were ponded up chemically millions of years ago, and we are releasing it now at a rapid rate. The earth has rarely if ever seen the rates of energy conversion and dissipation that our cultures have produced. It is not hard to see any of this, and the hard fact of it is sinking into the culture as we observe that our rapid use of energy has consequences on the whole planet, particularly on its atmosphere.

Omniculture

What I have been calling the Levolutionary Paradigm twists the normal perspectives that we have into new perspectives. The notion that what we call human cultures are really ecological communities containing humans in a dominant and leading role is one of these. A friend of mine has helped me to coin yet another new word that I think is most important because it will help us remember this new perspective.

The Omniculture is a futuristic goal. It is a utopian dream in which all of the planet's ecological parts; every species and their actual best interests, are truly represented by some kind of centralized decision making authority. I see this as a human-led development because we have the brains to accomplish it. The objective of the Omniculture would be to slow the plunder and waste of the planet, while keeping humans in the picture. Omniculture is a term that encompasses and implies all of the following unassailable facts:

1. Human-led culture has rapidly and recently become global. What was once many different, and smaller, human cultures, limited in general to the ecology of one part of the globe, is now unmistakably one global human-led culture, now with many sub-cultural parts. The parts are more than human history, they are the history of ecological regions, and the ways humans have managed ecological reality. They represent our history of cultures, and it is a history worthy of racial and cultural pride and self-respect. They got us to where we are, and where we are is in a process of Levolutionary aggregation among formerly separate cultures.

2. Humans have the power, the numbers, the intelligence, and the technology to make rapid changes to the ecological situation of the planet. As noted above, so-called human cultures are composed of many species that live in many regions of the

globe. It is not just the domesticated species either. Millions of organisms conduct functions we barely even know about. Regional differences result in different species being more important as food for humans than others, but global economics, international trade and commerce, have sewn us all together as a global, energy-transferring, ecosystem, and this includes our food, our petroleum, and every other form of energy.

3. Humans have invented two ways of encoding functional or holosystemic order that have transformed us into cultures based not on genetic relatedness alone, but based on ideas of relationship. Ideas are produced in human brains and this represents one new way of encoding information, the source of which is a mental evaluation of whether something perceived is good or bad for us. We perform natural selection on ideas as we think and make trial of words and deeds. Ideas come in all subjects, but one important class of them is about how to live with other people and with other species.

4. The second revolutionary evolutionary innovation of humans is that ideas formed in a human brain are able to be shared and communicated through the encoding of language and recorded ideas as in libraries and curricula. The curriculum of ecology exists, but this important science is not nearly where it would need to be for us, for Omniculture, to represent the interests of the planet's species.

These dual capacities of creating ideas and of transitioning them into language, policies, laws, constitutions, norms, and resulting behaviors are the foundation of how the Omniculture is now evolving the wholes that are biological organisms; people, plants, microbes, and animals we utterly depend upon, into the fit, holosystemic parts of a planetary ecological whole.

The planetary ecological whole has a population size of one, so far, and that is too small for comfort in a violently energetic universe. Humans, with our ideas and languages, are actually leading and profoundly affecting the evolutionary development of existing ecological communities and the Omniculture as a whole planetary ecosystem.

It is important to note that we are like babies who have just discovered their toes in this important matter. We have some knowledge, but we will need much deeper understanding, and we will have to learn how to incorporate the interests of all of the Omniculture's parts if we are to have a chance at maintaining our exalted position.

The progress of the ecological science is one thing, but the creation and enforcement of policies that represent the findings of ecological science is yet another. In this latter challenge, we are beyond infantile, as I have learned from first-hand experience. Consider even the challenges of long space missions in which we might hope to seed our Omniculture on another planet. Which species and abiotic processes would you take with you? What is the minimum complement? What principles are at work here? How can we reboot an ecosystem we want to live in tomorrow?

The human story is a story of evolutionary innovation and Levolutionary progress. Natural selection which drives both evolution and Levolution is, like it or not, at the root of it. Energy and its regularities, and indeed its purpose, that we have now carved and organized as the Thermodynamic Laws and the Laws of Order, are all wrapped up in the operation of natural selection.

THE THEORY OF LEVOLUTION

*The One is made up of All
Things, and All Things issue
from the One.*

– Heraclitus

THE LEVOLUTION REVOLUTION

We are now prepared to directly approach the important process I have
spent most of my life contemplating; the process that has built the 23
or so levels of organization observable in the universe. This chapter is a
bit lengthy because I wanted to record even the "nuances and wonders"
that have come to me in thinking about Levolution. Perspective, and a
bit of new knowledge are about all that is needed to understand it, but
it has created everything.

The process is called "Levolution" because each event of it results
in a new level of structural or material organization, and it is obviously
very closely related to evolution. By the end of this chapter the two
notions almost merge.

I do not mean "evolution" as commonly used by physicists, which
they use to imply a more general concept of "change over time". I mean
evolution, the process of adaptive change among holosystems by means
of Thermodynamic Natural Selection.

As a point of further clarification, however, I do not mean "evolution" exactly as it is viewed by biologists either. Reproduction and genetic encoding are not requirements of most events of natural selection, which may manifest as simple as "relative thermodynamic instabilities" among particles on the earliest levels of organization.

What I do mean is the more general principle of Universal Evolution, as the thermodynamic process of change described in the last chapter.

On the way to Levolution, I have gone through many perspectives about how to understand the relationship between it and evolution. I finally came to see the thermodynamic version of evolution as an important part of the process of Levolution. Evolution is the process of adaptive change. The difference of Levolution lies in the simple fact that it is the adaptive change of wholes into the parts of a larger whole.

For me it took the breakthrough of the holosystem model, form, or structure before I could also see that Thermodynamic Natural Selection and evolutionary change have a thermodynamically-defined target for the next larger whole. To be a good dissipative structure, or in the more common case, a good holosystem, is a purpose that is aligned with the cosmic project of entropy increase in the universe.

As you have seen, I bristle at the notion of self-organization, from which I hope to extricate this science of functional order-building. The external influence of Thermodynamic Natural Selection, and the external influence of energy resources are telling a different story. The clincher however is the external target, or the dynamic attractor state, which represents the pattern of energy flow in a holosystem. These elements are all missing from the current myth of self-organization, and these theoretical components are the heart of Levolution.

That said, to get here has also involved the matter of getting thermodynamics expanded with the relevant laws. I see no alternative to this course. For science to progress here, it must add the Maximum Entropy Production Law and the Dissipative Structure Law, or disprove them, or at least argue cogently against them. These are the two paths that converge on an understanding of Thermodynamic Natural Selection as yet another Law. Natural selection automatically yields

Universal Thermodynamic Evolution, and Levolution after that, is probably unavoidable.

Natural selection, that profound concept formerly reserved for deities, in its thermodynamic form will open the floodgates of existing evolution theory. I then see it all crowned by Levolution, the theory that explains the origins of the universe's diverse contents, the subjects of both natural and social sciences, as the universe's holosystems.

We finally get a unified Holarchy of Nature, a believable cosmology underpinned by thermodynamics instead of unexplained fluctuations, a logical physics that makes perfect sense, and a framework or paradigm that achieves a certain kind of intellectual gestalt.

The Levolution of Holosystems

Holosystems, our subject class of systems, include only those dissipative structures that are also composed of dissipative structures. In Levolution, Thermodynamic Natural Selection adapts a set of holosystems, often of several types, into the complementary parts of a larger whole, a new holosystem, and a new level of organization.

Levolution is a thermodynamic phenomenon, but while it is a universal principle, it is nonetheless a rare process. At some fundamental level it looks like the creative invention, by flowing and dissipating energy itself, of new scales and types of dissipative structures. Based on the number of levels of organization visible to us in all of nature, it appears to only have occurred about twenty-three unique times since the beginning.

My objective here is primarily to describe what is common about these processes that have caused, and is still causing, holosystems or dissipative structures to become aggregated into parts of a new larger whole that is also a holosystem. When Levolution happens, a new level of organization among the entities in the universe comes into existence. The process has built the nested structure we can see in the universe, characterized by sequential levels of organization among its entities.

I refer to the whole nested structure formed by viewing the various levels of organization all at once, as a "holarchy" after Arthur Koestler.

Figure 1 in Chapter 1 is a list of the levels of organization. Each row represents a level of holosystemic order, a whole, but also a part of the holosystem on the level below it.

The holarchic structure of the universe has resulted from the sequence over time in the origins of new holosystem types or in what might be called the chain of Levolutionary advance.

The holarchy can be segregated into line segments or Axes of Functional Order that seem to reflect the forms of energy. These Holarchic segments extend from the dissipating Singularity to quarks, nucleons, nuclei and to atoms on one axis, but then continues from molecules to planetary ecosystems, at least on earth. From a similar point near the atom, a Gravitational Axis of Functional Order branches off and extends from dust grains to galaxies. Each Axis is composed of links between part-whole dualities. What causes and drives this universal order-creating process? What are its characteristics?

There is no serious question about whether it has happened. While most scientists already see it, in a way, it has never been backed by a sound scientific theory of everything, an explanation for all the levels in terms of only a few natural laws. Even though Levolution is, by my reckoning, unquestionably the most important process in the universe, science has somehow left it for me to pull the relevant ideas together and tell you about it. We have all, by now, recognized the nested, structure of various types of systems in the cosmos, but what causes such systems to emerge as new levels of organization among the many dissipative systems that we observe?

Levolution operates when Thermodynamic Natural Selection, or the universal form of evolution, operates on populations of holosystems, and in rare cases on simple dissipative structures, to integrate them into a new and coherent whole. Differential survival among a set of holosystems is the universal tool of evolutionary change, and this is an important part, but only a part, of Levolution.

Even when one knows how to make changes, one still needs to know what changes to make, and nature would have the same challenge if it were not for the proposed Fourth Law of thermodynamics, the Law of Maximum Entropy Production. This new proposed thermodynamic law provides the teleology and the primary motivation for building new levels of holosystems on larger scales, and it gives wholes the motivation to become better adapted parts.

It is thermodynamically downhill to do these things. In other words, it is spontaneous, but that really doesn't tell us very much. Everything that really happens could be said to be thermodynamically spontaneous, and the word loses meaning outside of a contrived laboratory situation. I personally think that the word spontaneous, and confusion about its meaning, has led most scientists to the erroneous conclusion that things self-organize. They don't.

Levolution is a very general process, and at each of the levels it has created, the details differ a bit. They must differ because the scale change causes a change in the types of systems with which it is dealing. It may be similar to how quantum mechanics seems to match up with the very small, while general relativity matches up with the very large; and in between neither one seems very important or applicable. Levolution applies on all scales. It is generally a picture of functional self-similarity, but the details differ substantially.

Things change on the various scales, and the details of the process of Levolution change as well. What is common on all of the levels are the thermodynamics of each case of Levolution and a slightly less generalized pattern I will now relate.

Two Sub-Processes

The generalized process of Levolution might be said to reduce down to two sub-processes; differentiation and integration. However to get to the real crux of Levolution's operation, we must see these two modes as operating almost simultaneously and in concert.

Functional differentiation is not a random differentiation; it is a set of evolutionary changes among members some population of holosystems, and surprisingly it involves a direction in evolution *toward* the six generalized and complementary functions of any holosystem.

Now this idea might surprise biologists who understand evolution.

We have only recently gotten rid of the idea that there is such a thing as directionality or progress in earthly biological evolution, and it has been concluded that evolution is simply adaptation to changing circumstances, acting through random genetic variation and natural selection.

What gets us beyond that here is the entropic drive of thermodynamics. The six functions of holosystems, believe it or not, are all permanent niches, or one might say "state attractors" in the fitness landscape. It is as if the six universal functions that make up the patterns of a holosystem were carved in stone. The functional niches and corresponding roles found and occupied by evolving holosystems as they differentiate, are the very ones that are functional in moving energy downward as fast as possible by creation of a new dissipative structure. Let that profound coincidence sink in a bit.

The Root Causes of Levolution

To see this another way, the drivers and motivations for holosystems to come to occupy their particular functional thermodynamic niches is based on perspectives which mix the overarching thermodynamic view with the perspectives predicted for the real holosystems involved:

1. Start with an undifferentiated population of holosystems. Early in the process, before even the thought of a new level of holosystem is appropriate, a thermodynamic perspective on the initial unstructured population results in a view that the individuals are each doing all they can to move energy as fast as possible. Thus, one could sum up the total amount of energy that the unstructured population can flow through itself. Some members, however, are inherently and predictably hampered by entropic and situational factors. Being at the edge of the colony, being away from the edge, or being near a waste accumulation, may speed up or slow down the energy flow for certain individuals. *Inequality and variation just happen and may be assumed.*

2. From the thermodynamic perspective again, imposition of the holosystem pattern on the population would mean

differentiating the members to fill differing functional niches and roles. This would align the flows in a way that may be similar to the way they are aligned at a bathtub drain; it would segregate the opposed and divergent vectors, prevent collisions, and improve overall energy flow in the population. This would represent the order gained by developing energy-related, utility-like functions, but now call them functional, energy-flow-related niches, such as energy distribution and waste handling solutions. The benefits of these niches being filled would serve the whole population. *What is good for the individual holosystems of a population is also good for the population as a whole, and will increase total energy flow.*

3. Again from the timeless theoretical perspective of energy and thermodynamic principles alone, having those particular functional roles filled which lead to a holosystem will help to maximize entropy production by enabling a larger scale system. *Being bigger is favorable in terms of maximizing entropy production.*

4. Now from the perspective of the part-level holosystems, their own differentiation and occupation of niches is functional and beneficial for them too. It is like having a job. Having a role in the larger system will ensure that they are provided with energy, shelter, and a leg up in the struggle for existence. *Niches are evolutionary opportunities.*

5. From the developing perspective of the whole, the new higher order holosystem that is gradually coming into existence, the fact of the occupation of its functional niches by its parts is beyond functional; it is enabling. New systems on new levels of organization are creative solutions to the local problems and enable the whole to exist. The filling of its niches is what fills its needs as a holosystem. Their occupation composes it, completes it, and improves its coherence as a higher-order holosystem. *Holosystems in functional niches or roles enable higher order holosystems.*

These causal or motivating perspectives, or something very much like them, are probably operating in each instance of Levolution. This proposition is somewhat difficult to prove, and these motivating factors may be a small subset of the ones that may be found to be operating, but these intuitive ones are probably the fundamental ones.

Thermodynamic Isonomy

One of the more profound ideas to come from all this is the notion that energy seems to show a general disregard for differences between levels of organization. This is a result of a principle that I am calling "thermodynamic isonomy". Isonomy simply means equality under the same laws. The laws are the same regardless of level. What is good for individuals, is good for populations, and is also good for the next level up, and so on. What is good, of course, is maximal entropy production.

So, despite a lot of historical arguments by eminent biologists to the contrary, evolution into differentiated functional parts is directional and targeted in at least one sense. That sense being the Second Law of Order, the Law of the Holosystem Model. The holosystem model is a real energy flow model that is really important. It is carved in thermodynamic stone. This is the key to understanding Levolution, and Levolution cannot really be understood without it.

So, holosystems on one level evolve by natural selection into the differentiated parts of the holosystem on the next higher level. The process is brought about because there is a profound similarity between levels of organization in terms of the goal and purpose of energy in its flowing.

Equality under the thermodynamic laws, which I take to include the new Laws of Order, means that the same laws operate at every scale, at every level of holosystemic order. This kind of self-similarity is observed in other aspects of cosmology, but is only vaguely understood.

Indeed, there may be only one integrating tendency in nature, but it manifests itself in many ways. For Levolution to occur, two fundamental sub-processes must happen; differentiation of wholes into parts and integration of the parts into a whole. The result is differentiated parts integrated into a new whole, and the confusion about self-organization has surrounded the fact that the integration actually proceeds in concert with the differentiation. It almost seems like the new holosystem is organizing itself, but this is an illusion that is prevented by an energy-centric viewpoint.

Natural selection does not act randomly or alone. It is backed by an entropic teleology and a functional target that is an ideal form. When detractors from evolution talk about evolution being a matter of randomness or chance, they are referring to the random sources of initial variation. Natural selection acts on these variations, but it is never a matter of chance or randomness; it is quite the opposite. The functional target of natural selection is always the same; a holosystem.

There is an integrative aspect to **thermodynamic isonomy** that is at work in Levolution. At its base, the cause of Levolution is simple equality under thermodynamic law. This leads to both the inevitability and ubiquity that any good physicist will agree pertains to thermodynamic laws.

Specifically, in the process of Levolution, the integrating tendency exists because both holosystemic parts and whole holosystems are subject to the same thermodynamic laws, and these laws may now be seen to apply all the way down to the six universal functions of holosystems. As noted in Chapter 4, these functions are the capture, use, and dissipation of energy, and the "management" of these same three functions as conducted by the subject's parts, for a total of six functions. I am not trying to say that holosystems have only six parts. Their many parts simply, and at a minimum, all contribute to performing the six functions.

Given Thermodynamic Natural Selection as the mechanism of evolutionary change and thermodynamic isonomy surrounding the holosystemic target of change, Levolution operates to sculpt holosystems wherever it can; wherever constraints allow.

The Types of Levolution

We are now armed with the tool of understanding of how Levolution generally works, through sub-processes of differentiation and integration. Thus far, I have stressed that these processes are inextricably linked. Integration of the six functions of a holosystem is the goal and the model is essentially carved in stone. The energy is already flowing, so evolutionary differentiation need only produce the niche-occupying parts.

Sometimes these two aspects of Levolution are separated. There is more than one way of becoming integrated by Levolution. There are at least two ways. One way is differentiation in place from a population of more or less similar holosystems. The other way is for previously differentiated holosystems that have evolved separately into divergent entities, to come together as complementary parts.

Common sense tells us that the real difference here is simply where the process of differentiation occurs, it may happen internally with respect to the set of future parts, or externally, in which case they must be attracted or aggregated together later. Examples might clarify this somewhat.

Internal differentiation probably resulted in the earliest multicellular organisms, which probably are derived from colonies of similar and genetically related cells which differentiated *within the colony* to form perhaps an early sponge or a polyp. This situation describes what ecologists call *sympatric speciation*. Differentiation by biological evolution generally requires isolating mechanisms between parts of a population, but as I have noted above, these isolating mechanisms need not be geographic, and can potentially result from differences even within the differing regions of a colony. It is thus potentially our inability to see the isolating mechanisms at work within a hypothesized single colony of cells that causes the distinction here.

External differentiation is associated with the more traditional *allopatric speciation*, and in Levolutionary history this is what is thought to have occurred among prokaryotic cells. Their early differentiation led to various kinds of chemical energy specialists, including chloroplasts that can use sunlight, mitochondria that can convert many things to ATP which is the energy currency of the biological realm. The chemical template librarian that we now call a cell's nucleus, may also be an externally differentiated specialist. These are now all organelles; parts of modern cells. They have their own surrounding membranes, and even their own RNA.

After they differentiated, these energy specialists came together and found a home inside of a host cell. Their specialized functions became complementary functions of a holosystem and the modern cell has survived ever since. In biology we call this the *prokaryotic endosymbiont theory* of the origin of modern cells. It was first proposed by Lyn Margulis and has since become generally accepted. This mechanism explains how modern eukaryotic cells arose from an aggregation of previously differentiated, early, prokaryotic cells. Endosymbiosis is a perfect example of Levolution involving external differentiation.

In particle physics, external differentiation is the most common and understandable type. The attraction and connection process seems to be accomplished here in two different ways. The main method of integration among particles is by particle-produced, attractive forces. Attractive forces, like the strong force produced by gluons, the force of gravity hypothetically produced by gravitons, and the attraction between electromagnetically opposed charges, are all examples of attractive forces.

The strong force, for example, integrates quarks into nucleons, and gravity clearly integrates any matter within range of its force. Attractive electromagnetic charges account for the integration of stable atoms from electrons and nuclei, and also for the integration of chemical molecules from charged particles or ions.

I have come to believe, though it is radical and speculative, that one kind of particle can also attract another kind essentially by producing a form of energy that another particle consumes. In other words, I believe that every particle type has to consume some form of energy. It may easily be the case that one particle emits a form or a field of energy that another kind of particle needs to absorb.

In what can only be characterized as an ecological relationship, the consumer is behaviorally "attracted" or "drawn" to the producer of the subject form of energy, just as carnivores are drawn to herds of herbivores, and as trees are drawn upward toward the sun. Unconscious processes are all that I mean here, but note that evolution is not quite as unconscious as some would like. Evolution has designed some very elaborate structures in nature, and the behavior of holosystems is under Evolutionary control.

Particles may evolve to connect their flows of various forms of energy together with one consuming the entropic dissipation of the other. This is a good example of my admittedly strange ecological view of particle physics, but this instance is an important point that will come up again. I think this really has happened.

In this external differentiation mode of Levolution, previously differentiated entities, completely separate types, are attracted together by brute forces or by attracting them with food. As a result of the attraction, regardless of how it is accomplished, they may connect, they may find complementarity together, and Thermodynamic Natural Selection may favor them together. If favored, it will be because they are better adapted to their environment, i.e. they are more stable together, and/or they move energy faster together than separately.

The attraction and aggregation involved in external differentiation as a type of Levolution may be the result of a field of attraction, an attraction to some needed energy resource, or both. If parts are functionally complementary in this way, and fulfill the six universal functions of the holosystem model, the complementary parts may be stable together, and may exist together symbiotically, as the parts of a larger whole.

This kind of "integration through attraction and connection of energy flows" delivers some energy-ordering benefits. From the perspective of the shared environment and future whole, one captured quantum of energy may now flow through two holosystems in sequence, each differentiating it and degrading it in form.

This aspect of Levolution may essentially pull in other holosystems and string them together like an ecological trophic network. This would seem to keep energy flowing rapidly, within sequential dissipative structures, for a longer and more orderly path. The differing parts, and their produced or dissipated forms of energy output may speed up overall energy dissipation. If they come close to creating and integrating the six functions of a new holosystem, the integration will hold.

Perspectives on Levolution

The thermodynamic teleology of energy means that it is always open to a faster way to become degraded or dissipated, and it is always essentially ready to construct the six energy functions necessary to create a larger holosystem. This is the model behind the process that leads to natural selection and to differentiation of parts within a population.

The holosystem model is essentially what is adapting holosystems into the component parts of a larger one. A set of whole holosystems are adapted into their part roles by natural selection within their environment, but this environment is also viewable as the larger whole of the future. There must be naturally occurring variations for natural selection to act upon. Any source of variation could suffice, but the relevant ones are those related to the energy functions. The advanced evolutionary technologies of information-encoding and reproduction are not necessary for simpler holosystems, particularly ones that will recurrently form.

The sub-atomic particles, and the gravitational holosystems, are both characterized primarily by the "external differentiation" type of Levolution. At the higher levels of chemical reactions, cells, organisms, socio-cultural systems, and ecosystems we seem to observe the internal differentiating type, but my sense is that this is mainly a matter of being able to actually see the isolating mechanisms involved.

The holosystem model may be characterized as a state attractor, from dynamics and complexity theory; a state toward which spontaneous processes lead. Levolution may be viewed as either as a push from attractive parts pursuing their own interest in survival, or by a pull from the holosystem model. Thermodynamic Natural Selection is involved in both. Both types of Levolution operate similarly in that they seek a thermodynamic goal. Higher order holosystems are larger in scale and move more energy.

Levolution, operating through these two modes involving natural selection, is the process that has organized particles into atoms, atoms into molecules, and on earth, molecules into cells, and so on, up to the planetary ecosystem, which contains our culture and the ecological communities. Levolution also is evident in the way that gravity organizes molecules and dust into star systems, planets, and moons, and it organizes stars into galaxies.

Allow me to point out that Levolution, based as it is on the dissipative structure and holosystem models, is the *only thermodynamically legal source* of this kind of structural order known to science. Quantum theory has no believable horse in the race. Relativity sheds just a little light on it. String theory and Supersymmetry offer nothing useful here. Ecology; that's the ticket.

Levolution, like evolution, is really a matter of thermodynamic law. It is always going to be more fit to get bigger and move energy faster. It is always going to be possible for things to join together to produce a bigger, more powerful system. A new holosystem on a new level of organization is created because of the increased energy flow it causes or allows, which is to say that functional order creation is driven by the increased entropy it produces.

Levolution does not then apply some kind of thermodynamic lock, freezing the new holosystem in place. Instead, it is just the beginning. Natural selection will probably favor the new level, and it will probably take over its world. The parts will be continually refined and differentiated as needed by natural selection to fill all the available niches. Holosystems eventually achieve something like a near perfect evolutionary fit with their environment, but there is no guarantee or certainty. Sharks don't seem to get any form of cancer, but this may not mean that they have had sufficient time to achieve perfection.

The Holosystem Carved in Stone

The only "blueprint" or "design" for a holosystem is essentially etched into the laws of thermodynamics. It comes from the defining principles of holosystems themselves. This simple plan for an energy flowing system suggests the answers to many profound questions. It is a concept of thermodynamically created, systemic, energy-flowing order. It is holosystemic order; entropically functional order.

The holosystem model was initially derived through a theoretical aggregation of (1) General System Theory, (2) Dissipative Structure Theory, and (3) Levolutionary Holarchy Theory, and it obeys the two Entropy Laws.

Dissipative structures being pathways of flowing energy, are automatically subject to a precursor of natural selection, which is *energy pathway selection*, but it is rare to observe this. Only when we add the further qualification of "nestedness" and arrive at holosystems, does natural selection become common.

Biology has already shown us that natural selection can do some amazing things. We understand biological evolution fairly well. At least we thought we did. The Levolution Revolution, the Paradigm, is a flood of new awareness that comes with understanding these things:

- Natural selection is the resulting principle of a thermodynamic law favoring energy flow pathways delivering maximal energy dissipation. When those pathways are discrete dissipative structures or holosystems, the population level phenomenon resulting from their differential termination begins to explain the cosmic order.
- Dissipative structures exist to assist in the grand entropy dissipation project of the universe, which must dissipate the universe's energy potentials and degrade its forms of energy as rapidly as possible. These structures operate faster than simple energy conduction, by ordering the energy flows. Order speeds up entropy dissipation.

- Natural selection and "environmental constraints" share the function of natural law enforcement in biology, physics and chemistry. "Energy pathway selection" from the new Fourth Law, becomes "natural selection" if the pathways already exist as dissipative holosystems, and it becomes "constraints" on their becoming if they don't exist. The concept of "environmental constraints" exists and applies in physics and chemistry already; energetic constraints being most important here, so natural selection should be understood as both thermodynamic and universal.

- Evolution's thermodynamic function in the universe is to adapt holosystems to be more fit parts of a larger holosystemic whole, not merely to adapt systems to their environment. Environments are virtually always the inside of larger holosystems, which are also subject to natural selection. Biologists rarely take this perspective, but it is quite true.

- Levolution's thermodynamic function is to create new systems on new levels of organization, successively enlarging the size and scope of holosystems to move more and more energy. The series of holosystems that have been created up to the present represent the levels of the Holarchy of Nature. There are about twenty-three levels on three Axes of Functional Order.

- The holarchy of nature's holosystems results from Levolution, which creates each holosystem in the ecological position of a whole-part duality on a particular level of organization. Holosystems are dissipative wholes composed of dissipative structure parts, and they are generally a part of a still larger dissipative whole. Wholes must deliver energy to, and dissipate entropy from, their parts, but the parts use their energy, generally on behalf of the whole of which they are a part.

The overall picture of the universe obtained through the Levolution Paradigm is one of continuous unity, and continuous energy flow with a singular purpose. Its continual acceleration, in its mandated, downward, entropic direction, has resulted in the Levolution of one holarchy, sporting several lines of successively more ordered, energy-degrading and energy-dissipating systems.

These quantized units, the energy-manipulating holosystems, have become the main channels for all forms of energy flow in the universe today. Structurally they exist together in a nested holarchy, in which humankind and human cultures already have an important position. Human cultures are interfacing with the highest level of organization known, which happens to be the global ecosystem on the planet earth. The astronomical realm is larger, but we know of nothing more highly ordered than the ecosystems of which our cultures are parts, and the brains with which we hope to adapt faster.

MAKING SENSE OF LEVOLUTION

Levolution is driven by the second law, and the new fourth law of thermodynamics, and many of its details and sub-principles relate to the very fundamental properties of holosystems operating under these laws. The holosystem functions, for example, are universal.

Energy must always be headed downward in potential. Dispersive processes like conduction, form changes or degradation, and dissipation in space are usually how this occurs. Dissipative structures, which I take to include energy form-degrading structures, were an innovation that speeds up this process, usually by ordering the flows and encouraging increased energy capture. The basics were described by physical chemist, Ilya Prigogine in the 1970's.

The thermodynamics of structures and systems has opened up some new possibilities. The new Fourth Law says that given a choice of energy flow pathways, there is a selection principle at work. Among alternative potential pathways of energy flow, those pathways will be favored that maximize the rate of overall energy flow downward, and these pathways will receive and move the most energy. Based on my reading, I would give credit for this law to Rod Swenson (1988, 1989) who has articulately, and convincingly, championed it.

All dissipative structures represent and are composed of such pathways, and they are also discrete entities occurring in populations with varying traits or characteristics. This allows a restatement of the Maximum Entropy Production Principle as the law of Thermodynamic Natural Selection.

Holosystems, as discussed in an earlier chapter, are all dissipative structures whose parts are also dissipative structures. This nested structural arrangement is caused by Levolution, and is called a holarchy. The word is reminiscent of hierarchy, but it connotes a "nested wholeness".

Natural selection acting on holosystems is the mechanism behind both evolution, now understood in the universal sense, and Levolution. Far from being just the mechanism of Darwin's mechanism of biological evolution, natural selection, is a profound reality for every holosystem in the universe. That is why the major religious traditions include so many stories about catastrophes, and that is also why natural selection is a commonly attributed power of deities.

Levolution could loosely be thought of as a special case of Universal Evolution by Thermodynamic Natural Selection, but the process itself has a different function. The difference is that Levolution is reserved for the process that results in the origin of new types of holosystems on new, higher levels of organization. Natural selection acts similarly in both evolution and Levolution, but in Levolution it is more impressively "crafting" a novel and larger holosystem from available parts. In Universal Evolution, natural selection is adapting those parts as necessary to keep them adapted. Levolution has a loftier and more ambitious project. It has to build a new dissipative structure from scratch, and from available entities it can turn into parts.

Natural selection has the capacity to assist a set of holosystems in handling their own utilitarian challenges. Boundary challenges, entropic and material waste collection and elimination, and distribution of resources from the periphery to the inner regions, are just some of the ubiquitous challenges facing populations of holosystems. Just as civil engineers in a city might design an improved highway to alleviate congestion, natural selection can design, through its simple subtraction process based on variation, trial and error, thermodynamics, and time, what look like the utilities, transportation routes, organs, institutions, or simply the differentiated *holosystem functions* that serve the whole population of a whole's parts.

We know what these essential holosystem functions are generally, and that is the key. So does thermodynamics. In enforcing thermodynamics, natural selection, always favoring faster downward energy flow, carves the homogeneity away, and adapts the existing systems into better fitting parts, fulfilling those functions (the universal functions of holosystems). In the process, if we could see it, a holosystem forms before our eyes. Continued and constant differentiation performed by the ever-present possibility of natural selection to maintain, improve, and solidify benefits to the whole, eventually leads to a new system on a new level of organization. This is close to the essence of Levolution.

Levolution Defined

Levolution is the thermodynamically-driven, spontaneous process that creates a new and larger type of holosystem, on a new and higher level of organization, from existing dissipative structure parts.

Levolution is caused by isonomy of the thermodynamic laws, by a common thermodynamic teleology that exists as a matter of the entropy law, and is self-similar between both a new whole and a set of its prospective parts.

The convergent goal is accomplished by Thermodynamic Natural Selection, which operates on holosystems and selects for maximum entropy production. This differentiates and/or integrates a complete set of parts as the next higher-order whole.

There is an entropic teleology or a goal that is implied by the second law of thermodynamics. Energy has a mandate to find ever faster pathways downward in energy potential. This may be accomplished by formation of a larger dissipative structure. One way to do this is to take full advantage of the existing holosystems, and order them into parts of a new whole, a new level of organization, a larger dissipative structure, to increase the amount of downward energy flow possible.

The shared thermodynamic mission and principles of entropy increase, and the shared holosystem solution-set existing on both levels, of parts and wholes, is key. Energy capture and entropic waste elimination, for example, occur on all levels of organization. Energy distribution to parts, entropic waste elimination from parts will also be challenges of the new whole. It is a matter of holosystem thermodynamics that the whole will align some of the individual contributions of its parts into the useful energy output of the whole. The whole will generally be seen to surpass the former disorganized population in terms of energy flow rate.

A population of holosystems inherently shares the holosystemic challenges and the holosystemic solutions for producing order and maintaining it. Such a population will be driven by natural selection to develop the required differentiations and complementary aggregations to conduct the universal functions of holosystems and so become a higher order dissipative structure. Sufficient energy must be available, and other constraints must allow this.

In other words, a population of dissipative structures is naturally on the path to become a new higher order holosystem, if constraints will allow it.

Levolution happens in events that look like a coming together, an aggregation, a collapse, coalescence, or union of dissipative structure parts. The key however is that the parts, in coalescing, have also organized into a new holosystem that moves energy faster than the parts would have individually. Achieving differentiation and complementarity among the parts, a predictable necessity for organizing more complex holosystems, is accomplished by Thermodynamic Natural Selection, which is non-random, and acts in alliance with, and in fulfillment of, the thermodynamic goals of entropy.

Levolution is thus primarily the result of *isonomy under the thermodynamic laws*. It happens because the applicability of thermodynamic laws is universal, on all levels of organization, and includes the powerful principle of natural selection, the teleology of entropic dissipation, and the six universal energy moving functions of the holosystem model.

Levolution and Evolution

Levolution is a special case of thermodynamic evolution, acting through natural selection, on a set of associated, but independently dissipative, holosystems. While evolution and Levolution both adapt wholes into more fit parts of larger wholes, Levolution adapts its subjects of selection into the parts of a new holosystem; one that was not there before. Once Levolution creates it, evolution keeps it adapted.

While the Levolution Paradigm gives us the perspective change from viewing the system-environment relationship to the part-whole relationship, that is not the difference between Evolution and Levolution. Both Evolution and Levolution are truly adapting the parts of wholes, and our thinking has been slightly incorrect about evolution in this regard. The only real distinction between the two processes is the *novelty of the level of organization* involved in cases of Levolution.

Biological evolution should properly be viewed as the process that is continually adapting organisms to be the fit parts of ecological communities, and similarly, is continually adapting cells to be either the fit parts of multi-celled organisms, or remain as fit and essential parts of the ecosystem at large.

Therefore, for anyone who cares about really understanding biological evolution, it must now viewed through the lens of the Levolution Paradigm. We simply need to recognize that the way we have thought of evolution since Darwin, is slightly deficient. Natural selection is not merely adapting systems to their environment. It is adapting systems that are dissipative structure parts into the more fit parts of a larger, higher order holosystem, which is the ecosystem.

Distinguishing Biological Evolution and Biological Levolution

Darwin's theory of evolution by natural selection was about the biological process that creates new species and produces change within species, and creates new types of earthly life. Levolution is a closely related theory in that its mechanism is also natural selection, but it is about the simultaneous creation of new types plus a new level of organization. Levolution has happened in biology at least four times. It happened when prokaryotes became eukaryotes, when single-celled organisms became multi-celled organisms, when many organisms became structured populations or societies. And it happened at least once more as the many species became ecological communities and systems.

Astute readers may note that there is a slight problem here with the pattern. Actually every organism, even single celled ones, were in an ecological situation from their inception. The problem is that we are using ecology to mean two things. It is both the science that applies to all system-environment relationships, and it is the modifier we use to describe the biological level of organization in which many species live together as a functional system. Another problem is that the population size of the planetary ecosystems is one.

In addition to, and certainly between these "events" of Levolution, the more common process of evolution kept all the systems well-adapted and in existence to transfer as much energy as possible through the ecosystem.

Evolution results in a diversity of well-adapted types, as functional, energy transferring parts in a functioning system, in this case an ecological community. The amount of diversity is a function of the complexity of the system, but Levolution is not characterized by a continuous increase in the complexity of holosystems over time. Complexity must simply match the challenge of building a holosystem that works on the next higher level and scale of order. One could argue that the chemical reaction sets of a modern cell are more complex than an organism like us, with only a few organ systems, yet we Levolved and evolved much later.

Ecological communities and larger units are complex. Diversity provides increased probability that the needs of holosystems can be met through flows up and down through the biological segment of the holarchy, and ultimately helps to dissipate it into outer space.

Revisiting the Theater Analogy

G. Evelyn Hutchinson, a much revered, well-known ecologist, used an analogy here. Evolution and ecology are like a theater. There is an evolutionary play, and it is happening on the ecological stage. Animals evolve and change over time, but at all times they make perfect sense in the "current" ecological setting. This is a good analogy to understand how ecology and evolution operate together, but it does not really address the roles of the actors.

Hutchinson also gave us the *multidimensional niche space*, a way of looking at ecological niches in terms of measures of fitness across multiple environmental variables, and the bounds in these parameters within which an organism may survive. Charles Elton had earlier invented the concept of the ecological niche. His concept was based on the notion that each species had a specific role in the ecosystem. The Eltonian concept of the *role-based niche* brings us back to Hutchinson's analogy of the theater, now armed with the concept that the various organisms are playing necessary, functional, evolutionary roles on the ecological stage, while they are subject to natural selection and the change it causes. In this emerging picture, the plot is always similar.

Now that we know that natural selection is a matter of thermodynamic law, and that there are necessary holosystemic functions that are the same at all levels, the theater analogy may get more detailed. The theatre analogy becomes this updated synthesis of this situation.

Organisms are playing thermodynamic, holosystemic roles in the continuous, universal play entitled "Entropic Dissipation". The same play is always being performed, but the actors come and go. The various roles in the play are descending energy flow pathways also known as ecological niches. Various actors may take on the energy–transferring roles in the play, and the roles always need to be performed, as the plot never changes.

Because the roles are essentially arranged as trophic levels in the flow of energy, they tend to connect end to end. Some roles convert sunlight into plants, for example, and others, the herbivores, consume the energy in the plants. They use it for their purposes, but also dissipate and degrade it. The actors may be yanked off the stage by a stage manager with a hooked stick (natural selection) if they don't perform their roles properly, or if competing actors can perform them better.

Ecology is the study of biological systems and how they interact with their environment. This is a very broad field, not only because there are millions of types of organisms, but also because there are a myriad of interactions possible. As the field has matured, certain principles have emerged. Ecosystems, food webs, metabolism, even migrations, the general struggle for existence, and evolutionary adaptation to resource scarcity are all energy-related processes.

Ecology's major finding has been less seriously described as "Everything is connected to everything else." That is very true. Ecosystems are a web or a network formed by natural selection among holosystems, to capture, use, dissipate, and degrade the solar energy that falls on the earth. If you map the flow of energy through an organism, you will have gained an understanding of the fundamental structure, and the internal organs of that organism. The same is true of ecosystems. Trophic levels are an important aspect of their structure. The web-like nature of ecology is based on the fact that energy flows through it. Energy is what has driven life to organize, flow in networks of holosystems, and energy is what drives evolution.

Levolution creates new kinds of holosystems on new levels of organization for only one reason; to get bigger. The growth of a new level is thermodynamically spontaneous. It organizes a whole set of individual holosystems, expanding the amount of energy that gets ordered into a new system. Expanding also the range over which it might explore, the speed with which it can move, etc.

Evolution maintains an adaptive fit of every type of holosystem to its environment, but that environment should always be recognized as a whole holosystem. The natural selection process that operates is operating on the parts of that whole to keep them well-adapted to their roles.

Directionality in Levolution and Evolution

As we know, great care must be taken in discussing the directionality of evolution. It has always been safest to say that it is always simply toward "being well-adapted". The course of evolution is indeterminate in terms of predicting what types of things may evolve. Trends are not unknown, but it is difficult to know if they will continue.

This is still the case with regard to Levolution, but there is a directionality and it is upward. The next level will be predictably composed through a "coming together" of holosystems that exist now, possibly as the highest current level of organization.

There is still much that is indeterminate and unpredictable, but Levolution can be said to have a discernible upward direction. We humans may easily overestimate our own role in this play, but the stage manager will do whatever is needed.

Selection for Community-like Functions

Isonomy, equal subjection to the thermodynamic laws, including that of Maximum Entropy Production, and the principles of special thermodynamic systems like dissipative structures, and energy-degrading holosystems, gives us the theoretical underpinning for natural selection among holosystems. Venturing outside of biology with this concept in hand should be educational, and it is. Consider gravity.

Gravitational holosystems are probably the simplest of all, and their situation is free from any complexity, prejudice, or possible emotion. It is a logical speculation, and it is a new hypothesis here that gravitational holosystems feed on space's energy, which they dissipate into what I will, for now, call "degraded space". The useful work that they do is the production of gravity itself, and this apparent force is the work done through their space consumption. Gravity is a deficit or rarefaction in the smooth distribution of space around gravitational objects, which Einstein insightfully called a "curvature" of spacetime.

To see how gravitational holosystems conform to our Levolutionary concepts, consider first that it's a jungle out there. Specks of matter, clumps of matter, and huge collapsed masses of matter are flying around at a range of velocities through spacetime, and each one of them exerts the familiar force of gravity. As they encounter each other their forces combine into a mutually attractive force between them. In such encounters there are what we could call winners and losers. Winners are the gravitational holosystems favored with continued independent existence by natural selection, and the losers either get collapsed into the winners, becoming a part of them, or they fly entirely out of the frame of reference. Either way, the losers are gone, the parameters of their fitness were simply their position and angular momentum.

The collapsed matter in gravitational holosystems is ultimately a population of gravitons. As the probable components of quarks, gravitons may be structured into quarks, nucleons, nuclei, atoms, ions, or molecules, but there could also have been gravity long before these levels came into existence. The massive bodies or gravitational holosystems that collapsed into the central mass were losers because they lost their independent existence. Their matter now, however, is contributing to the force of the central mass. Gravitational order is simple, radially symmetrical, and simply achieved by a "coming together" as a new force-producing holosystem. Each time another mass is added in a gravitational collapse, an instance of Levolution is occurring, but as noted above, it is not a "counted" instance.

Gravitational Levolution is a very common event. No complexity exists to require even a care about the material being aggregated. Only mass and density may have a bearing on its response to the gravitational force. Once associated, cooperation is unquestioned, and escape is impossible. Utility-like solutions to entropically caused challenges of the population are not in evidence because entropy in the gravitational holosystem is simply the emission of degraded spacetime to the outside of the system.

Assuming the boundary of a gravitational system is its Jane's radius, the outer limit of gravitational effect on stationary particles, the entropic emission would take place far, far away. However, there is a caveat about this if the gravitons are structured into atoms of ordinary matter.

In nucleons, three graviton clusters that we know as quarks are held together by gluons and the attractive Strong nuclear force that they produce. Gluons are situated among quarks because they consume the degraded spacetime that quarks produce, and use it to produce the Strong nuclear force. It is likely that gravitational holosystems made of ordinary matter will thus absorb much of the degraded spacetime, which is the entropic product of gravitational holosystems. For this reason, we might not expect to see the entropic product of degraded spacetime, when gravity acts on normal matter.

When gravity is produced by gravitons that are unassociated with normal matter (in the absence of gluons), we might well see such an entropic product as degraded spacetime being ejected by gravitational bodies. In Chapter 9 I will reveal how this might impact our understanding of Cosmology.

The Inter-level Ecology of Holosystems

The reality of Levolution and the primary evidence for it, which is the whole Holarchy of Nature, provides us with a view of each whole-part duality in which we can often simultaneously see (1) the types of holosystems that are its parts, (2) the type of holosystem that it is, and (3) the type of holosystem that represents a larger whole, of which it is a part, and this is commonly where its reason for being, its primary energy-moving function, becomes visible.

The ecological stage is set and the roles to be played have been rehearsed many times. Viewing the situation of any particular whole-part duality, energy must come from the outside, from the whole, from the environment. While it may still have to be collected, one of the aspects of the bargain struck in Levolution is that the new larger whole will become the new free-agent in the more random environment, and the wholes that are now to be parts, will be "served" its energy by the whole. Wholes take care of their parts, and in return, parts will do the work that will be managed by, and attributed to, the whole.

Along the Holarchic segments of the Axes of Functional Order, energy flows downward from wholes to parts, just as it flows from any environment to the systems within it, but energy also flows back upward as the useful work of parts which "animate" the wholes, and make each level possible.

This situation is very telling with regard to understanding the ecological and holosystemic frame of reference, and how natural selection operates at all levels simultaneously to continuously maximize the overall flow of energy in the universe. The main point I want to convey here is that with this new view of a part-system in a whole-environment represents a new view of ecology itself. We have a lot of work to do to update the science of ecology in view of Levolution.

Critters are not out there in nature just because they can live in some niche. They have a thermodynamic purpose. They have a thermodynamic role to play as a holosystem part within a larger holosystemic whole.

Selection Pressure from Above and Below

Natural selection is operating everywhere at the same time, and all the time. This thermodynamic process may act on a specific type of whole-part duality in many different ways. Many millions of years of biological evolution have undoubtedly been blasted into oblivion, and perhaps even totally wasted, by meteorites and comets colliding with the earth. Selection pressure in earth's ecosystem may come from predators or from competitors, but it may also come from outer space.

It may also come from inner space. A disease of individual cells may result in the death of whole organisms, and plagues may affect, if not obliterate, entire societies. As the parts are being Levolved into a new whole, the emergence of the whole is also the emergence of a new way for natural selection to operate, and a new subject for it to operate upon. Through no fault of the part, a whole may be destroyed and the other parts along with it. Similarly, through no real fault or poor adaptation of the whole, a lower level calamity can take it out of the play. There is still plenty of randomness at the edges of evolution and Levolution. Everything is indeterminate to a degree, and this is what Levolution is really up against. Creating order out of chaos is its very function.

The Academic Challenges of Levolution

Levolution is a process that has been happening since right after the big bang, and it has created virtually every durable, naturally occurring, energetic system type in our universe, and that includes most of the particles, the atoms, the chemistry, all the protists, plants and animals, us as individual humans, our culture, and our earth's ecosystem.

There is little or no argument about whether something like Levolution has occurred. How else would the systems of the universe, the subjects of scientific disciplines, be arranged in levels the way they are? There will be few objections about the sequence of production of new systems on new levels of organization emerging in time. Atoms, molecules, cells, organisms, and societies, simply by their components and structure, demonstrate the phenomenon of Levolution over time.

Levolution's existence as an operating process could not really be seriously in question, but the process of its level-building has never been adequately and accurately described. The surprise is that Levolution is the default standard, if not the exclusive, thermodynamic modus operandi for undertaking its primary project in the universe, increasing entropy. It is a direct and easily visible aspect of, or strategy of, cosmological development.

It doesn't take a rocket scientist, and I am not one, but the new perspectives that work so well are ecological. We have the science at hand for the process of Levolution to be described very scientifically. Even though this book is not very formal, the science included here can easily be made formal. In general, however, I would feel secure in debating with non-believers on the basis of what is here.

There are some confusing aspects of the newly revealed reality, especially if you have already learned thermodynamics, or taught it. The history of this subject shows that the science of order needs to advance. Some aspects of thermodynamics need to be "unlearned". Entropy is not the universal tendency toward disorder. Entropy and its production is the motivation and drive to produce functional order. The energy is becoming more degraded and dissipated all the time, but the functional order is becoming richer, deeper, and larger in scale at the same time.

The biggest problem I see, as one who wants Levolution to become accepted science, is primarily that it lies largely within the discipline of physics. Physics in general is quite obviously indifferent to advances in thermodynamics. The important laws discovered by Onsager, Swenson, and Prigogine, came nearly a half century ago, and they have yet to be codified.

Physics seems a bit hostile to biology, from whence many of Levolution's principles have come. Physicists are well-known to also be somewhat arrogant about their discipline. One can see all kinds of confusion and turmoil ahead. Levolution will be disruptive to physics. That is fine with me. It needs a shake-up in my opinion.

As I am not willing to simply toss the ball of Levolution over to physics, it is clear that garnering support for it will be important in my future. Please register any such support at Levolution.com. There is a lot of experimental and theoretical work to do to nail down the Levolution Paradigm. Levolution will not leave such things to chance. Ideas, if they are true, will survive. That simple statement says everything I need to say.

CHAPTER 9
THE WONDERS OF LEVOLUTION

The learning of many things teacheth not understanding, else would it have taught Hesiod and Pythagoras, and again Xenophanes and Hekataios.
— Heraclitus

THE REALMS OF LEVOLUTION

The process of Levolution in the abstract is now under our belts. The logical case has been made and the supporting observations generally noted. What I want to do now is take a different slant on aspects of cosmology, considered very speculatively but with the benefit of Levolution.

For me, Levolution was my perspective before I learned much about particle physics or cosmology. Tantalized by Lee Smolin's *Life of the Cosmos*, I decided to try revisionist ecological cosmology as a concept. Everything has to eat something in this framework. Particles are not waves, they are holosystems. My objective in this final chapter now has changed. I do not seek to convince you of anything, but to speculate within the framework and show the new direction in which Levolutionary Cosmology seems to lead. The next book is centered on this view, but I wanted to provide a preview here.

Particle Levolution

In the Levolution of sub-atomic particles, Levolution is a matter of aggregation based on attractive forms of energy. Take a look at Figure 6, which has been taken from the upcoming Levolutionary Cosmology book to help summarize the situation here. This graphic portrays the first levels of organization in the universe as a speculative look at the Levolutionary development of the particles.

The photon probably formed as a dissipative structure and the gluon evolved from photons by means of natural selection. The never observed graviton is then a composite of photons and gluons, held together by the strong force. A version of this speculative particle is the dark graviton, responsible for dark matter. I apologize for the suspense, but that story is in the next book. Gravitons themselves aggregate by the attraction of both gravity, which they produce, and the residual strong force of the gluons, into the simplest quarks. These are the cores of the Leptons, including the electron.

The electron and the simple up quarks combine perhaps into the down quark. We have now made it up to the realm of the less speculative knowledge. We know that up and down quarks combine into neutrons and protons, collectively known as nucleons. The nuclei actually get smashed together within stars and supernovae and form the various elements. This is counted as a single event of Levolution when they stabilize into atoms through combination with electrons.

Gravitational Levolution

Through the Levolution Paradigm we have an understanding of gravitational holosystems and the operation of gravitational natural selection. We have the observations of astronomy, and so we know that the bodies in space are indeed found to be arranged in levels of organization. We call the levels, in order of increasing mass, dust grains, planetesimals, moons, planets, stars, and galaxies. There are other names for variants of these. They are all approximately spherical or orbital. The interesting exception about the Levolution of the gravitational bodies is that Levolution did not make them form in the sequence of increasing mass.

Figure 6—The Levolution of Particles

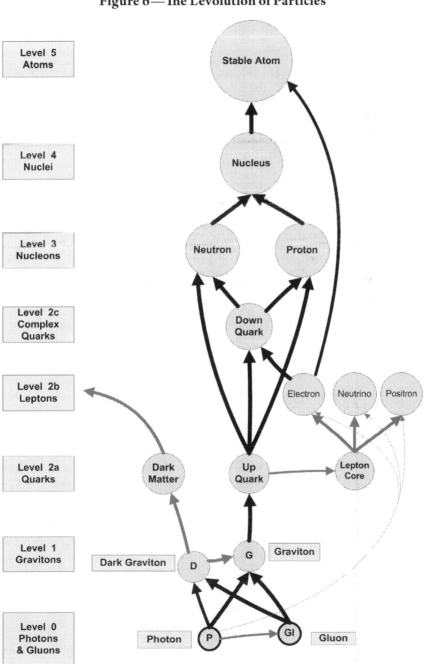

Particle Levolution seems to show a definite mass increase over time, and the story of electromagnetic Levolution will show this stepwise enlargement in size as well.

The difference here is that gravity is much more dependent upon distance or space than it is with time. Gravity's work will happen based on position and momentum of objects, any objects, and it will happen inexorably, almost determinately. Time is of the essence wherever energy is concerned, but gravity is so simple and determinate that it is not the course of time that will mainly determine a gravitational object's size. It is its initial mass and location, and the positions and masses of its neighboring objects.

Also, many galaxies formed before many stars because gravity began working on those galaxies before it worked on those stars. In fact, the Large Scale Structure of the universe, which is not mainly gravitational, formed before there was gravity. This structure was formed mainly by the expanding voids, which pushed the matter into what we now see as filaments and clusters.

Electromagnetic Levolution

The universe is generally electromagnetic. It began as the singularity, but the first thing it did was to divide into two electromagnetic poles as the Duality. The early Particle Levolution story here is speculation based on my new considerations of space. It involves energy forms which begin with, and are sequentially derived from, this primordial polarity or electromagnetism.

The weak force is already known to be related to electromagnetism. Space, the strong force, and gravity are as well, but that Theory of Everything is in another book. The other thing to note is that the completion of the atom with electrons, and everything that is chemistry is clearly electromagnetic.

Less clearly, but just as true, everything biological, including thinking with a brain is also electromagnetic in nature. The sunlight that powers the earth's ecosystem is electromagnetic. The computer on which I write this is electromagnetic, and all the motion of the culture is caused by muscles, motors, or engines, that reduce to electromagnetism, chemical or otherwise. The Electromagnetic Axis of Functional Order is a long one. Gravity looks like an afterthought by comparison.

Let's look briefly at the part of the electromagnetic Levolution that is most mysterious to people; the origins of life. Figure 7 shows the succession of levels of organization between the atoms and the modern cell. Atoms combine through chemical reactions into molecules. Reactions, where the energy actually flows to make molecules, form families of specific reactions that catalyze each other, and form cycles. At least they are thought to have done this on the earth.

Given the magic of chemistry turning into living cells, it is not hard at all to see how prokaryotic cells invaded each other and became modern cells.

While not shown, we know that modern cells became colonies, became organisms, and became societies, and became modern human cultures expanding across the planet.

All the while, of course, the complex chemistry and the biology was all being carried out within the earth's ecosystem. The ecosystem, like the galaxy, was there first, but it is always changing in response to its components. There is some complexity to deal with here because there is only one known planetary ecosystem. It may be a complex system, but it is also a captive system, tied as it is to a gravitational object. It is also not a part of a population, so if it meets with its demise, it may not really be a case of natural selection. All this being known, and said, I think the earth's ecosystem is still the top holosystem; a hybrid between a gravitational and an electromagnetic one.

Ideational Levolution

We know that something new is going on right now on earth that is changing things rapidly. It is our expanding global culture. Human cultures started as genetically based tribes, and gradually they have become what they are now. They are idea-based social, economic, and ecological structures that involve a host of other species and utilize

Figure 7—The Levolution of Cells

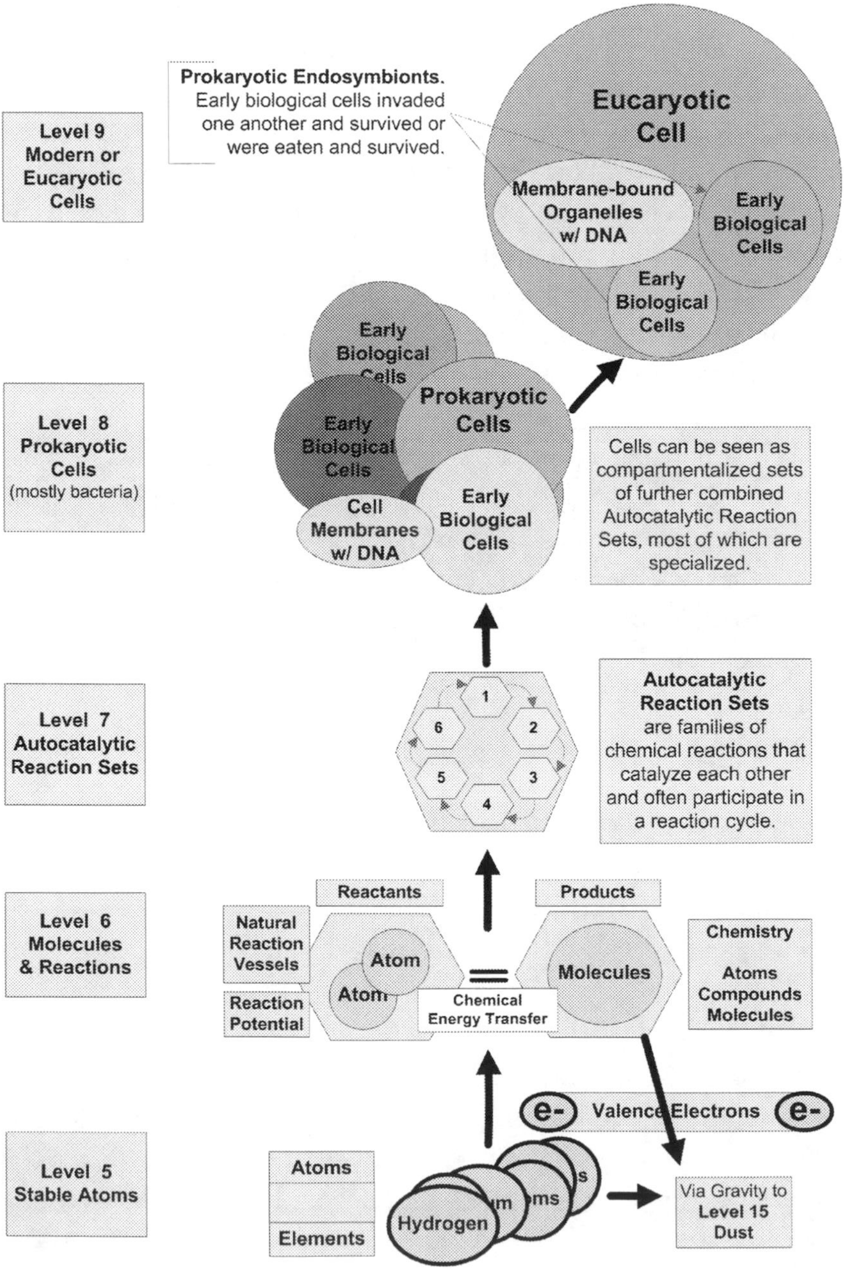

many of the natural resources of the planet. Laws and policies govern much of what happens, and this is an improvement over the lack of these that prevailed before. We are constrained by the social order, and that is why we have been able to survive this long.

Have a look at Figure 4 again. This diagram shows all of Levolution on its three major Axes of Functional Order. The Electromagnetic Axis continues out to its current terminus, which is labeled as the planetary ecosystem. But note that each of the last few levels of organization on that axis have something new about them. They each sport an informational innovation; a form of encoded functional order. This observation has to have an explanation.

My explanation is that one way to speed up energy flow is to speed up evolution, Levolution, and the growth of functional order. Encoding the functional order, as we have already noted in discussing biology, speeds up and improves Universal Evolution by Thermodynamic Natural Selection. Basing cultures on easily modified and communicated ideas and memes does the same thing. We can change our laws if needed in a very short time. It will take another book to tell this tale, but it will pick up the trail here.

THE POSITION OF LEVOLUTION

The case I want to try to make here is that the holosystem pattern is a real thermodynamic feature of a real subclass of dissipative structures. As biologists would say, there is a true *homology*, not a mere *analogy*, between the levels of organization of holosystems. In evolutionary biology, homology refers to the reason why some characteristic is found in common between an organism and a possible ancestor. Here I will not use it to describe a character trait, but to describe the Levolutionary process of new holosystem creation on the various levels. The contention is that the very same thermodynamic process, not just a similar process, is at work on all levels.

Figure 8—The Levolution of Ecosystems

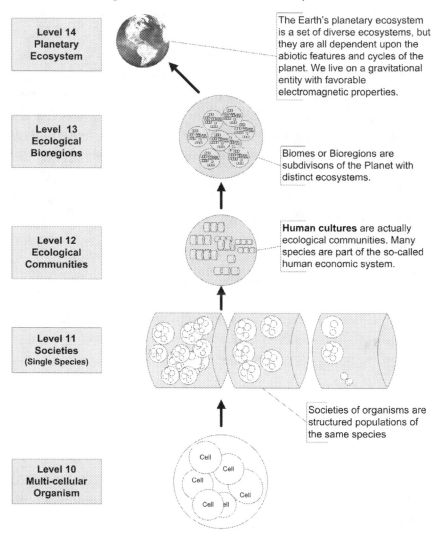

**Level 14
Planetary
Ecosystem**

The Earth's planetary ecosystem is a set of diverse ecosystems, but they are all dependent upon the abiotic features and cycles of the planet. We live on a gravitational entity with favorable electromagnetic properties.

**Level 13
Ecological
Bioregions**

Biomes or Bioregions are subdivisons of the Planet with distinct ecosystems.

**Level 12
Ecological
Communities**

Human cultures are actually ecological communities. Many species are part of the so-called human economic system.

**Level 11
Societies
(Single Species)**

Societies of organisms are structured populations of the same species

**Level 10
Multi-cellular
Organism**

Characteristics or characters can evolve more than once, but when they are derived through inheritance from an ancestor they are homologous characters, and if the character arose independently, it is merely analogous. What I am saying here is that the holosystems of the universe are all related, and the similarities of the holosystem structural model, a universal thermodynamic model from which they were all patterned, makes them all thermodynamic homologues.

Holosystems are a feature of the universe that arose through the regularities of thermodynamics. The same schematic pattern of energy flow that provides one level with thermodynamic stability will also provide the next level with the same benefit. Levolution steers the universe's development. It is a very tight and complete theory of order creation consistent with the new thermodynamics and the new Laws of Functional Order.

The following are some other interesting observations or conclusions I have come to learn about Levolution. They are not categorized in any way:

1. Holosystem are a modified dissipative structure pattern among pathways of energy flow. The energy flow connects between the levels of organization and its rate of energy flow increases when a new holosystem emerges. *Entropically functional order produced by all dissipative structures is the universal basis for co-operation.* It is actually the enabler of simple things like pooling efforts to capture resources, like working together toward a common goal, and like installing utilities that improve functioning for all the members of a population. Talcott Parsons even based his perspectives on Sociology on this very same functionalism.

2. One can look at the process simplistically as a simple selecting, sorting, and organizing of energy pathways such that opposing forces are segregated and energy-wasteful collisions are avoided, but when a whole's parts are dissipative structures, certain further assumptions can be made. A relatively increased flow of energy will already be flowing in the individual patterns that reflect energy capture, use, and dissipation on the lower level. The pattern is already half established before Levolution really starts, and so creating collective and

distributive utilities to integrate them is the primary challenge overcome by Levolution and the holosystem pattern.

3. Adrian Bejan's recently described **Constructal Law,** what I know of it, sounds real, functional, and I can see that it could "plug in" here to explain an important sub-aspect of Levolution. Levolution, at least on several high levels of organization, involves building collective and distributive flows of energy and materials to serve a population of individual parts. Energy distribution and coordination of the work output of parts are both essentially what Bejan calls "flow networks". It is a phenomenon of the same type of morphodynamics which gave rise to dissipative structures. I think it will someday become integrated as a commonly observed part of holosystem development. Constructal Law seems to explain the growth and development of the *energy-related utilities* that characterize the process that integrates the parts in wholes. It seems like it may have Thermodynamic Natural Selection as an understated central mechanism.

4. The notion that entropically functional order increases, or even maximizes, energy flow is a foundational principle of dissipative structures. It is also true of holosystems, of course, since they are dissipative structures. This beneficial effect on energy flow rate may be overtly recognized as the *functional order principle* of dissipative structures.

5. Holosystemic or functional order is a means of increasing entropy production and is an important part of the universal project to dissipate energy potentials. Such order thus enhances the thermodynamic fitness of holosystems essentially because *functional order shares the interests of energy itself.*

6. H.T. Odum's Maximum Power Principle, which he derived from the study of ecosystems, is close, but it does not seem as correct as the Maximum Entropy Production Principle, even in the ecological application. Maximum Power, or work per unit of time, is not the same as maximal energy throughput,

but maximum entropy production is the same. Energy scarcity may lead to efficiency-maximizing considerations, while energy surplus may leads to power-maximizing considerations. If you can choose, you won't lose. The more general statement of the choice is the MEP, and this can apply over both the short- and long-term of survival. This is equivalent to a simple view of energy flow as maximized within operable constraints that I think even Odum was probably seeking.

7. The very notion of dissipative structures and holosystems, as discrete systems, adds a twist to thermodynamics. Because discrete systems can transfer no energy if they perish, they acquire a *thermodynamic imperative to survive if possible*. Survival in energy scarce environments may lead to a new temporary thermodynamic teleology related to efficiency, which is not in entropy's general teleology at all. Similarly, and perhaps more generally, survival may occasionally require the maximizing of power, the rate of doing useful work. Some holosystems are even consciously aware of things like survival, usefulness, efficiency, and power, while disordered energy, in view of the conservation law, would not be the source of such a perspective.

8. Holosystemic order always implies the six universal energy functions of holosystems, represented at the level of a whole by (1) energy capture, (2) energy use, and (3) energy dissipation. It is represented on the level of its parts as (4) energy distribution to parts, (5) management of the entropy of parts, and (6) management of the collective energy use, the commonweal, or the work output of parts.

9. Holosystemic order creation is a downhill or spontaneous thermodynamic process, but the work of natural selection may not be as instantaneous as it appears to be in chemistry, hydrodynamics, or computer simulations. The rate of biological Levolution provides a clear example of why the word "spontaneous" only applies through its thermodynamic meaning, as the "downward" direction of a reaction, not its more generally

understood or "normal" meaning as something like "self-ac-complished" "quick" or "unplanned".

10. Order creation through Levolution involves natural selection and thermodynamic fitness. In its differentiating mode, a pop-ulation of similar wholes solve their entropy-related challenges (waste accumulations, energy and material distribution needs, and boundary condition maintenance needs) by differentiating into new types that participate in the conduct of utility-like, energy-transferring, ecological, or housekeeping functions for the population. In its integrating mode, disparate parts from any source are attracted by an energy source, connected by complementarity, and subsumed into an integrated holosystem because the aggregation is favored by Thermodynamic Natural Selection.

11. Both Universal Evolution and Levolution select from among the various possible pathways of energy. The "natural selection of discrete downward energy flow pathways" is a phrase that describes most of what goes on in the universe. I think this must be related to time, because it is this which keeps time from going backwards.

12. Levolution is driven by energy's drive to maximize entropy production and energy's rush toward lower potentials. The se-lection of alternative paths of energy flow, by energy itself, is based on maximizing the entropy production rate, and here it should be noted that energy *use* results in entropy production just as surely as the ejection of unusable energy does. These are merely perspectives from the vantage point of an individual holosystem.

13. The discrete pathways, which is to say the structures and systems, that increase energy flow the most will most likely be more fit than others. This is essentially the foundational statement of Thermodynamic Natural Selection and Universal Evolution theory, which occurs in the light of the Law of Maximum Entropy Production.

14. A new system's Levolutionary emergence can be seen as the result of a simple congruence that exists between (a) the utilitarian energy needs of a set of holosystems, and (b) the universal functions required of a higher order holosystem. Both parts and wholes need to capture and distribute energy, eliminate wastes, maintain boundaries, preserve order, etc. and that has a profound implication in that it leads directly to Levolution.

The Lucky Congruence

A simple congruence exists between (a) the utilitarian energy needs of a set of holosystems, and (b) the six universal functions required of a higher order holosystem.

A population of things with individual needs represents, to any higher level observer, a whole set of similar needs that adds up to one big need. A solution here is likely to be larger in scope and scale. Holosystems are all energetically similar, and this notion of energetic self-similarity allows them to scale up by way of Levolution.

This surprising functional congruence, this homology of energy functions between levels of organization, is based on the similarities between all holosystems. *This thermodynamic congruence allows Levolution to operate with two loci of natural selection.* One locus is selecting and differentiating parts based on their part-role performance, and the other locus is differentiating those very same parts based on the performance of the whole of which they are a part. Selection acting on the system one level up, will effectively select against its parts as well.

These are represented by two phases of evolutionary differentiation based on these two layers of natural selection. The first is the functionally adaptive differentiation based the on utilitarian needs of the population of parts, and the latter is the fine tuning, increased coherence, and, in some cases, centralization of control achieved by the new whole over its parts, in adapting to its own environment.

The utility-like solutions relate to solving predictable population level problems related to the universal energy functions. The utilities are such things as energy distribution and waste elimination, which arise because even uniform members of a population will be differing distances from edges, and differing distances from waste accumulations. These environmental differences will translate into differentiation of wholes into diverse parts.

The differentiations involved in the development of these population utilities will be adaptive because every improvement will incrementally increase the energy flow through the population.

The later phases of this process are really just the increasing coherence of the whole, until we achieve the so-called "emergence" of the holosystem in the conscious mind of the observer.

The patterned flows exist, of course, whether or not observed, but a characteristic of Levolutionary emergence is that holosystems may be cloaked in the scale-induced invisibility that comes with the inability to perceive the smaller systems. Darwin came before ecosystems had been described as such, so he can be forgiven for not seeing that the ecosystem as a whole which adapts its organismal parts. It is true nonetheless.

The Evolution of Levolution

My own understanding of Levolution changed in the middle of writing this book. Initially, I thought I understood that I must present two main theories; (1) Universal Thermodynamic Evolution and (2) Levolution. The first theory would serve as support for the second because evolutionary change via natural selection is actually most of the mechanism of Levolution as well. But a theory of Universal Thermodynamic Evolution did not yet exist. This jewel has been on the table since Herbert Spencer, but has not ever been elevated to the status of Law, and has never been seen as the universal mechanism of adaptive changes in functional order that it really is. This theory also would have to be clarified and expounded.

My further thinking, and discovery of Rod Swenson's insights about Maximum Entropy Production as a selection principle, led me to see that the same process, natural selection, drove both of the above processes. Both phenomena are essentially proven real by the observations we can make with our mind regarding the existence and nested structure of the systems in nature, or as I say, by the holosystems. These are the real subjects and results of these processes.

So I accepted that these two processes were actually the same thing, and the only difference was that one process (Levolution) created new levels of order, and the other (Evolution) simply maintained the systems on a given level. Evolution preserves the holosystems in a state that matches the challenges and opportunities of their environment by adapting the systems when environments change. Darwinian style evolution, only now made universal, was relegated to a maintenance role, adapting and differentiating holosystems as necessary to keep up with changing, or newly discovered environments.

Levolution seriously upstages evolution by producing all the new levels of functional order in the universe. The realization of something new then slowly emerged in my mind. Biological evolution, updated by the current observation of ecosystems as holosystems shows us organism types or species serving, and evolving, as obvious functioning parts of an ecosystem. This is also appropriately considered the adaptation of parts within a whole.

Biological evolution is not typically viewed as the continuation of the process that caused adaptation of wholes into parts. Evolution has always, even since the development of ecology, been viewed simply and primarily as the adaptation of systems to their environment. *It is much more informative to understand evolution as the process that continues to adapt organisms into the fit parts of an ecological whole.* It is information that may even save us from ourselves.

The perspective I have come to then, is that all of biology, and indeed everyone who seeks truth, must essentially change their perspective on evolution. The Levolution Paradigm shows that we have always had it slightly incorrect. Since Darwin, we have failed to see that organisms in environments are actually organismal parts within ecological wholes. We did not properly see the Holarchy of Nature as a structural edifice built by Levolution, and we did not see natural selection underlying it all, so we did not have the clues.

With this additional perspective, we can see that evolution, like

Levolution, is also really based on the phenomena of parts and wholes. The only difference really is that Levolution is the part of the process that originates new wholes, while evolution is the same set of processes serving to maintain those new wholes in an adapted state to maximize energy flow.

Evolution and Levolution are, in a way, parts the very same process. With a view of the levels of organization in the Holarchy of Nature, it is simply that Levolution creates, and Evolution maintains and adapts holosystems. Both processes are parts of the single process of degrading and dissipating energy to the maximum by using the strategy of creating nested levels of dissipative structures.

Or, and this is important, you might state the interesting alternative perspective that energy, dissipating and degrading at the maximum rate is building, maintaining, and energizing a self-serving, entropically functional order within it. It is possible that the functional order being built has a hidden or unrevealed destiny of its own. It's certainly getting smarter.

The process of Levolution reaches, at each moment in time, some state along the Axes of Functional Order. At any given time, nature can be imagined as a snapshot.

The best picture of the highest order, electromagnetically-powered holosystem we can imagine is a snapshot of earth's Planetary Ecosystem. We know that it produced us, it contains us, and it supports us. We like it a lot. The only disturbing thing is that this point of Levolutionary advance, this result of 13.7 billion years of functional order's development, leads to a picture of earth's complicated and multi-faceted nature, being almost "led" into its uncertain future, by the ideas operating in this culture.

Holosystems Are Real

First, I thought there were two theories; Evolution and Levolution, then I thought there was only one big story and it was really about natural selection. Natural selection when acting upon a population of systems may differentiate previously undifferentiated material systems, and when diverse systems come together through attraction, natural selection may favor the combination over separateness. Levolution is always dealing with thermodynamically similar parts. It is always ready to build a thermodynamically similar, larger whole to move energy faster.

There is a blueprint, similar to Figure 3, the Holosystem Model. It is not written or encoded anywhere except in the Laws of Functional Order, in the genes of biological organisms, and it will now be, for the first time, found in a library of human culture. The model is maintained all the way up and down the holarchy, in the corpus of every single holosystem.

Holosystems are not just a descriptive device or a theoretical construct. Holosystems are a class of things displaying a special thermodynamic pattern of energy flow, but we must put almost all natural systems into the category, so its usefulness is mainly in distinguishing them from simple dissipative structures, and in recognizing the universality of the pattern.

Holosystems seem like a very real thermodynamic pattern of energy flow and dissipation that are more durable than convection cells, whirlpools, and tornadoes. Holosystems are made of self-similar holosystems, and they are all subject to natural selection and have some inherent evolutionary flexibility.

If things aren't quite right, they adjust. They grow into it, if there is empty space, and they are attracted to their energy sources. They abhor a vacuum. Nature, even in its mere obedience to its own laws, seems very, very wise to me.

A Functional Difference

When Levolution happens, it creates a new, higher level of organization; one that was not there before at all. It is like a new invention. Evolution involves the same exact process of natural selection, but the heavy lifting of getting to a new holosystem, a new level, has already been done.

The functional role of evolution is to speed up energy flow by keeping everything well-adapted to its position and role as a part in a holosystem, in its environment.

The functional role of Levolution is slightly different. It is to make energy flow even faster by adapting the currently highest-order systems, into the parts of an even higher-order system, which will also be larger and have a larger range and scope.

Natural selection is still operating down at the part level (and below), and now it is also operating at the newly created, higher level as well. As long as energy flows, natural selection favoring the steepest, widest pathways of energy flow will occur. If there is a new, higher-order system, it will still be immediately subject to this principle of natural selection among holosystems, but the flows involved will be steeper and wider.

Levolution marks the events in the history of the universe where each holosystem type, which are dissipative structures with parts that are also dissipative structures, are first legitimately viewable as the parts of some larger whole. After the parts are there operating as a whole (the product of Levolution) they are kept in an adapted state by Darwinian-style, but universal evolution. Levolution is a rare event that has only happened about 23 unique times. Evolution, on the other hand, is always operating, and has differentiated millions and millions of separate system types.

Basic Levolution

The universe as we know it is the result of the growth in holosystemic or entropically functional order. The actual process that builds holosystems and creates functional order is Levolution. It is thermodynamically spontaneous, or downhill, but that does not explain it.

Levolution proceeds as a process of evolution of wholes into parts by Thermodynamic Natural Selection, but the environment to which holosystems adapt is an environment that shares many aspects in common with the evolving holosystems themselves. What makes Levolution work is that correspondence, a coinciding of the energy-moving functions between holosystems on successive levels of organization.

Hints of something like Universal Thermodynamic Evolution and Levolution have existed for a long time. Ancient philosophers have remarked about them. The Great Chain of Being, from the time of Plotinus in ancient Alexandria, was an early and related conception of a stratified order inherent in the universe.

With our current, interdisciplinary understanding of science, almost everyone has already recognized that for things to exist in a nested structure on levels of organization, those levels must have formed sequentially. Levolution is, without a doubt in my mind, the distinct process that did it.

These two theories, Universal Evolution by Thermodynamic Natural Selection on one hand, and Levolution on the other, may be viewed as the merger of thermodynamics and ecology, and the extension of evolution theory to all the systems of the universe. The two processes really do unite our view of physics, nature, and humanity in a profound way.

The change of perspective here affects all of the scientific disciplines, both the natural and social sciences. There is no fundamental separation, either in nature or in our sciences, between the operations of nature, humans, and cultures. They all capture and dissipate energy as they do work that is useful to energy itself in its project to dissipate.

As this sinks in, many of the sciences come into view in a new way, as studies of energy flowing in very similar systems. It is still amazing to me that most of these systems obey the same natural laws and share so much in common. Similarities among naturally occurring dissipative systems, combined with similarities among their environmental challenges and situations, leads to the broad and universal applicability of the Levolution Paradigm.

My objective has been to point out the general principles involved in Levolution, and show how they apply to the various system types found in the universe. The result is the outline of a new paradigm that explains one reasonably simple process that operates to create, and adapt as needed, virtually everything in the universe.

We now know generally how energy creates the functional order of the cosmos. The story has been told without reductionism, at least without old-style reductionism, without math, without probability waves, and without deities, unless Energy counts. The Levolution Paradigm yields a perspective that connects the structures of matter with the functional patterns in energy flow. This energy-centrism makes a big difference.

My Next Book

Through the lens of the Levolution Paradigm, the universe looks different than before. In my nearly completed, next installment of this Levolution Series, I will describe a completely new Levolutionary Cosmology based on it. It is a mind-expanding view to be sure.

The next book involves a change in mental gears. Up to this point, I have sounded fairly certain in my perspectives; even to the point of codifying proposed natural laws. From here forward, we are embarking on an adventure in which we are armed with these proposed Laws and new perspectives, but we are using them to think, play, and creatively speculate. Our subjects will be the barely known aspects of the universe, and our goal will be to advance science in new and productive directions that we cannot even see without the Levolution Paradigm.

The Levolution Paradigm will enlighten the Big Dissipation, space creation, photons, gravitons, the Levolution of particles, and the mysteries of cosmology. A Theory of Everything will emerge and both dark energy and dark matter will be explained by a speculative tale of ecological sounding processes at work in the living universe.

From here forward, we want to use the Levolution Paradigm to help fill in, and in some cases rectify, our knowledge and perhaps see if the more detailed universe, as observed through astronomy, cosmology, and particle physics seems like the same one whose Laws are herein codified. We will at least pick the low hanging fruit.

GLOSSARY

Autocatalytic Reaction Set
An Autocatalytic Reaction Set is a family or set of chemical reactions with the property that some reactions catalyze or promote other reactions. Cellular metabolism is an example of such a set.

Boson
A type of massless, energy-carrying particle, which in the Standard Model of Particle Physics is the type responsible for energy transactions using the Gauge Theory. Photons, for example, are bosons that carry electromagnetism, and gluons are bosons that carry the strong force. Levolutionary cosmology views these bosons as simply producing these forces because of their structure and the character of their components.

Degradive Structure
Technically, the same as a dissipative structure, but in this case emphasis is placed on the transformation of one form of energy to a new degraded form with lower potential, rather than simply dispersing or dissipating energy in space.

Dissipative Structure
A theoretical morphodynamic structure in flows of energy (often represented by fluids in motion) that uses incoming energy from its surroundings to create and maintain its structure of energy flows, or its

functional order, and also dissipate or degrade more energy, or produce more entropy, than before it formed. It represents the only way known to produce functional order.

Eco-Community
Shortened from Ecological Community. In ecology, the concept of multiple types or species of populations of systems living together in an energy-transferring relationship that tends to sustain continued survival of all the types, subject to natural selection.

Eidos
In the philosophy of Plato in the 3rd century B.C.E. the world we know is an imperfect representation of an entirely separate world of Ideals and Ideal Forms that exist in a non-physical, purely intelligible realm. The Eidos was Plato's Master Ideal Form, from which all the others are derived. Herein, an association is considered between Plato's Eidos and the dissipative structure of thermodynamics, which is an ideal form or energy flow pattern, of great importance to the thesis of Levolution. All the holosystems of the universe which are the subjects of Levolution, share this ideal energy flow schematic, form, or structure.

Electromagnetism
An important and primordial form of energy, which may have emerged from the degradive and dissipative polarization of the Singularity into positive and negative poles, followed immediately by their physical separation which is also the creation of space, and its energy content. Electromagnetism here represents an entire Axis of Functional Order, involving structures ranging from stable atoms to planetary ecosystems.

Electron
A particle of the Lepton class in the Standard Model, characterized by its negative electromagnetic charge. The electron is an important orbiting constituent of stable atoms that is generally attracted to the positive charge of the nucleus. Populations of orbiting electrons typically balance the number of protons in the nucleus to produce a stable atom. They are very energetic and mobile in conductive solid material

lattices, but their motion is not self-propulsive, and requires opposed charges. Such motion is known to us as electric current.

Endosymbiont

A composite of the terms *endo* (inside) and *symbiosis* (living together in a mutually beneficial way) used to describe the situation hypothesized in the Prokaryotic Endosymbiont Theory of Lyn Margulis. The origin of modern (eucaryotic) cells is considered likely to have resulted from the invasion or ingestion of one or more prokaryotic cells by another one, followed by realization of their mutual benefits, and their internal mutualistic survival. The ancestors of these ingested cells are visible today as membrane-bound organelles within modern cells.

Entropy Law

The Second Law of Thermodynamics. This important law sets the direction of energy flow and states that energy always flows downward in terms of its quantitative potential, strength, or concentration. This is a profound goal-directedness and represents the universe's teleology. It means that energy is always degrading and dissipating, but it does not mean that order is always decreasing.

Entropic Drive

Given the Entropy Law, the Maximum Entropy Production (MEP) Law, and the fact that energy is universally required for anything to happen, it is simply a matter of logic, and a manner of speaking, to recognize that the objective or motivation for anything that happens in the universe is to produce entropy at a maximal rate. This is Entropic Drive. It is essentially another way of stating the MEP Law.

Functional Order

The meaning of Order in general is the notion of a pattern, sequence, structure, or form, and the functionality implied here is "entropic functionality", which in the light of the MEP Law, is anything that helps energy flow downhill as fast as possible, within constraints. As it turns out, the patterns of a dissipative structure are entropically functional, and so a conception of the "Dissipative Structure Order" or "Holosystemic

Order" is the same idea. It is a generalized, thermodynamically-defined pattern that is as close as science has ever been to the Platonic Eidos. Contrast this with thermal-kinetic order.

Holarchy

The Holarchy or Holarchic Structure is the pattern of the universe's various types of holosystems in terms of their levels of organization. It is essentially a conceptual classification in which each type of holosystem is a functional part-whole duality. Each entity on each level is a whole, but it is composed of parts, and it is a part of a larger whole.

The environment of any given subject is actually the internal milieu of the entity above it in the holarchy. This means that all of the important energetic phenomena of dissipative structures are involved in this structure. Energy allocation to any such system comes from its environment, and all of the entropy produced by a given system is released to its environment. Even the work done by a subject holosystem is expended to the benefit of the larger, higher order system, represented by the environment of that subject system.

Simply put, the Holarchy of Nature is also the nested structure of the universe as a whole. It is being produced by the Monodyne (the singular flow) of Energy, as it degrades through a cascade of energy forms, each one more degraded, dispersed, or dissipated, and with less potential than the one before. These various differentiated forms of energy are used by the holarchy's material systems, the holosystems, in creating the Cosmic Order.

Holosystem

A holosystem is a dissipative structure, a special thermodynamic form, which is also composed of parts that are also dissipative structures. The relevance of this recursive requirement is that it limits the holosystem subjects to only those dissipative structures that have formed through the process of Levolution. This large subset of all the dissipative structures, which otherwise would include convection cells and tornados, create a clear record of the parts, wholes, and flowing energy, through which Levolution has operated to produce all of the material entities of the universe.

Information

Information is an innovation in Levolution's advance that is produced by encoding functional order. The detailed pattern of actual energy flow in a holosystem is encoded into a form or an implicate order that is (a) less clearly physical, (b) more ethereal, (c) easier to manipulate, disperse, and communicate, and (d) less expensive to move and store (in terms of energy and space). Encoding implies a direct correspondence between the actual and the represented functional order, and this correspondence is captured in things like base pair patterns in genetics, or character order in words.

The classic definition of information by Claude Shannon is not about the content of information at all, but is about its encoding and transfer or communication. Shannon's Information Theory concerns how easily information may be encoded and communicated in a message, how many symbols or bits are needed to convey a message, and how difficult it is to eliminate uncertainty from information.

Information Entropy

In Information Theory, which has developed along Shannon's lines, and was derived from Boltzmann's quantitative notions of thermal-kinetic order, a property of communications called "Information Entropy" is a measure of the amount of uncertainty that exists in a communicated message.

Levolution

Levolution is the process by which energy in the universe has created sequential new types of holosystems, each on new, successively higher, levels of organization. It is a process driven by Thermodynamic Natural Selection acting to evolve the parts of a new holosystem structure from populations of existing holosystems. One might call the dissipative structure pattern the target of Levolution here, but the holosystem as the "target" of the natural selection is better because it already implies that the new level of organization will be built from the existing populations of entities. This seems to always be the case, but it may simply be because existing populations of holosystems are where the most energy is flowing.

Monodyne Cascade

All of the energy in the universe arrived at once and has been flowing ever since. It differentiates into new forms as it degrades, and it dissipates across expanding space as well. It is still energy, and is in fact, still the same energy. It has a long history, and is reused but ultimately degraded. So, because of its Singularity or initial oneness, and its conservation, it could be called a Monodyne, because it is a single batch of energy. The history of the Monodyne is captured in the picture of the Monodyne Cascade, a stepwise, downward- flowing degradation and dissipation through sequentially degraded forms.

Niche

The ecological niche is some aspect or part of an environment in which a particular species of thing has adapted to survive. There are two notions of the niche, the Eltonian or role-based niche, and the Hutchinsonian conceptual hyperspace niche. The role-based niche would say that each species occupies some important functional role within an ecological community, but the functions are not always clear, or clearly divided, among species. The more recent theory, the Hutchinsonian niche, is a multi-dimensional, purely conceptual, hyper-volume with dimensions of environmental variables, enclosing the region within which the species can live. The Eltonian, or functional role-based niche concept is somewhat more useful in conceptualizing and understanding dissipative structures and holosystems.

Nucleon

The nucleons are classified as Hadrons, which are composites of three quarks. Nucleons include the positively charged proton, and the neutral neutron. Nucleons combine in various numbers of protons and neutrons to form the nuclei of the various elements.

Nucleus
The nucleus is a composite of protons and neutrons, and due to the charges of these are typically positively charged. They are formed in stars and supernovae which produce pressures sufficient to overcome weak force repulsion and allow permanent strong force attraction between the nucleons. There are about 98 naturally occurring types of elemental nuclei, and they are numbered according to the number of protons they have.

Omniculture
Omniculture is a utopian future state of the planet earth, in which the entire set of functions required for the continued and/or balanced operation of the planet's various abiotic and biotic cycles and functions, are managed. In this futuristic world, the important processes that keep us alive are ensured, guided, supported, and protected by its dependent members. These members are led by, and their interests are both known and truly represented by, futuristic, advanced, and very wise humans. The science of humans, by this time, has discovered all the required knowledge and technologies to assess and preserve the functional ecology and the planetary cycles of the earth, and this advanced science has been used to develop sound policies and a supra-ecological, Omnicultural order among the large-scale activities on the planet. Omniculture is a far off goal, or objective, but is based on the notion that human cultures, which are already ecological in their true nature, will need to lead this effort, or submit to control by other forms of natural selection.

Order
Order is simply a pattern or arrangement in some set of things, and the type of order seems to depend either on the type of the patterns or on the type of the things. There are the following types of order that might relate to physics: numerical, chronological, positional, thermal-kinetic, and entropically functional. The last two types relate to thermodynamics directly, and the last type named actually makes energy flow faster,

but it is only found in dissipative structures, and its name is herein shortened to functional order.

Photon

The Photon is probably the first particle, and formed as a dissipative structure, a pattern in which electric charges and their trajectories were ordered to achieve faster energy flow. The result was a wave-particle duality that self-propels at light speed, essentially by pulling itself along on a straight line in the field of space. In terms of the three-function model of the dissipative structure, the photon consumes space, works to pull the line and self-propel, and dissipates a degraded form of space.

Quark

The quark is, speculatively, a composite of gravitons and this gravity-based structure forms the core of other particles. Most leptons probably have a core made of quarks. Up quarks are the simplest. Down quarks are a composite of an up quark and an electron. Three quarks always compose the nucleons. Protons have two up- and one down-quark. Neutrons have one up- and two down-quarks.

Reaction Potential

In chemistry, the holosystems include the chemical molecules, but also the naturally occurring "reaction systems" in which they react. At a higher level, bordering on biological, the holosystems are entire sets of related reaction systems that catalyze and promote each other. In all chemical reactions, the reaction potential is the amount of energy available to make the reactions proceed from reactants to products. The reaction potential is usually discussed as the primary form of chemical energy, and is based primarily on the affinities of the valence electrons of the atoms involved.

Singularity

The energy of the universe is thought to have begun at a point, with an entity called the Singularity. The Singularity subject to thermodynamics would have been very unstable and would be required by thermodynamic laws to divide and dissipate as fast as possible.

Space Energy

The electromagnetic energy of the universe was set up by the two opposing poles of the Duality, which followed the Singularity. Space was formed by and in the discharge of energy that followed, where the two halves of the energy in the universe flowed toward each other and discharged to create space between them. Space, like light, is a phenomenon of electromagnetism, but space actually received most of the energy of the Singularity.

Spacetime

A mathematical treatment of space will usually begin with consideration of its structure in three dimensions. Time is another phenomenon of energy (irreversible energy flow) that is often structured into a linear dimension for mathematical purposes. The dimensions of space and time are both human-created ideational constructs, but they do enable mathematics. Einstein showed that they can be put together into four-dimensional spacetime, and that this multi-dimensional hyperspace would capture all events of motion, but this notion of spacetime is modified by the presence of mass or gravity which deforms it.

Relativity says that time moves slightly faster for the outer edge of a spinning phonograph record on a turntable, than for the inner part. The important idea is that space and time are both malleable and stretchable. Euclidian geometry is our normal geometry, but it does not explain all of the phenomena of real space and time that emerge when it is known that photons always travels at a constant velocity in a given medium. General Relativity implies that only a stretchable, or relative, concept of space and time can exist.

Teleology (Entropic)

Teleology is Aristotle's Fourth Cause. It is the goal or purpose for which something is done. In science, since Francis Bacon, teleology has been frowned upon due to an association with the "divine purpose" which we may not know. However, it was argued that there is indeed a natural teleology by Liebnitz, and it is actually inescapable that the Entropy Law (with or without the Maximum Entropy Production Law), represents an energy-based teleology of degradation and dissipation, a teleology that even physicists must admit applies to the entire universe.

Thermal-Kinetic Order

Thermal-kinetic order is here defined as the type described by Ludwig Boltzmann, and is essentially "spatial togetherness" as that property relates to the probability of various patterns of dispersion among particles of an ideal gas in a box. Maximal dispersion and dissipation is the most probable state, and is the equilibrium state, which in dynamics is a state attractor. In thermodynamics, equilibrium is the lowest level of energy potential spontaneously achievable. This low-energy type of order increases with coolness, or the relative absence of heat, and is greatest when the ideal gas transforms into condensed matter, or crystalizes, as a phase change. Contrast with Functional Order.

Thermodynamic Ecology

Thermodynamic Ecology is the science of the relationships between the thermodynamically important forms, which is to say the dissipative structures and holosystems, and all aspects of their function. This includes a thorough consideration of their environment, from which their energy comes, and to which their entropy goes, and the fact that their environment is most likely the interior of a higher-order structure. It is a direct generalization of the biological concept of ecology. Thermodynamic Ecology goes along with the recognition of Thermodynamic Natural Selection as a universally applicable phenomenon, which actually underlies all of the parameters acted upon by biological and other realms of natural selection.

Thermodynamic Fitness

Derived logically from the quantitative ecological variable "fitness", usually denoted as W, "Thermodynamic Fitness" is still the opposite of Selection Pressure, S, and so $W = 1 - S$, but this more general and universal notion is based on the new Law of Maximum Entropy Production (MEP) and the Law of Thermodynamic Natural Selection (TNS). Thermodynamic Fitness retains the ecological notion of "adaptedness" but it recognizes that underlying all of the parameters of natural selection is a more fundamental and thermodynamic set of priorities and a corresponding layer of deeper causation.

Universal Thermodynamic Evolution

Adapted from the biological theory of change among plants and animals that was Darwin's evolution, Universal Evolution is the more general theory and is applicable to any discrete population of variable entities. Remove some members based on energy pathway selection, as manifested in Thermodynamic Natural Selection, and the resulting mix is changed. Universal Evolution always means adaptive change through the mechanism of Thermodynamic Natural Selection. It occurs on all of the levels of organization in the universe.

Universal Functions

The universal functions of dissipative structures are the (1) capture (2) use, and (3) dissipation of energy. The universal functions of holosystems are these also, plus the management functions that go along with having dissipative parts with their own universal needs. In brief, they are (4) energy allocation and distribution to parts, (5) management of the work output of parts, and (6) management of the entropy production of parts.

REFERENCES

Bak, Per. 1996. *How Nature Works: the science of self-organized criticality.* Springer-Verlag New York Inc.

Brooks, Daniel R. and E.O. Wiley. 1988. *Evolution as Entropy: Toward a Unified Theory of Biology.* Second Edition. University of Chicago Press.

Burnet, John. 1930. *Early Greek Philosophy.* Fourth Edition. Adam and Charles Black. London. *First published 1892.*

Dawkins, Richard. 1989. *The Selfish Gene* (2 ed.), Oxford University Press. ISBN 0-19-286092-5

Deacon, Terrance W. 2012. *Incomplete Nature.* W.W. Norton & Company. New York.

Einstein, Albert. 2001, 1916. *Relativity: the Special and General Theory.* Translated by Robert W. Lawson. Dover Publications Inc. Mineola, NY.

Elton, Charles. 1933. The Ecology of Animals. Methuen London.

Kauffman, Stuart A. 1993. *The Origins of Order.* Oxford University Press, New York.

Kauffman, Stuart A. 1995. *At Home in the Universe.* Oxford University Press, New York.

Koestler, Arthur. 1967. *The Ghost in the Machine.* The Macmillan Company, New York.

Laszlo, Ervin. 1987. *Evolution: The Grand Synthesis.* Shambhala Publications, Inc. Boston.

Lotka, Alfred J. 1922. Natural selection as a physical principle. *Proceedings of the National Academy of Sciences*. Vol. 8, pp 151–4.

Margulis, Lynn. 1970. *Origin of Eukaryotic Cells*. Yale University Press.

Morowitz, Harold. 2002. *The Emergence of Everything*. Oxford University Press.

Odum, H. T. 1995. Self-Organization and Maximum Empower, in C.A.S.Hall (ed.) *Maximum Power: The Ideas and Applications of H.T.Odum*, Colorado University Press, Colorado.

Oldershaw, R. L. 1989. Self-Similar Cosmological Model: Introduction and Empirical Tests. *International Journal of Theoretical Physics, Vol. 28, No. 6, 669-694, 1989.*

Prigogine, Ilya. 1980. *From Being To Becoming: Time and Complexity in the Physical Sciences*. W.H. Freeman and Company, San Francisco, CA.

Salthe, S.N. 2010. Maximum Power and Maximum Entropy Production: Finalities In Nature. In *Cosmos and History: The Journal of Natural and Social Philosophy,* **Vol. 6, No 1 (2010).**

Sayre, K. 1986. Intentionality and information processing: An alternative model for cognitive science. *Behavioral and Brain Sciences* 9, 121-166.

Schroedinger, E. 1945. *What is Life?* The Macmillan Company, New York.

Shannon, Claude E. (July/October 1948). "A Mathematical Theory of Communication" *Bell System Technical Journal* 27 (3): 379–423.

Smolin, Lee. 1997. *The Life of the Universe*. Oxford University Press.

Smolin, Lee. 2007. *The Trouble with Physics: the rise of String Theory, the fall of a science, and what comes next*. Houghton Mifflin Company. Boston and New York.

Smolin, Lee. 2013. *Time Reborn*. Houghton Mifflin Company. Boston and New York.

Swenson, Rod. 1988. Emergence and the principle of maximum entropy production: Multi-level system theory, evolution, and non-equilibrium thermodynamics. *Proceedings of the 32nd*

Annual Meeting of the International Society for General Systems Research, 32.

Swenson, Rod. 1989. Emergent attractors and the law of maximum entropy production: foundations to a theory of general evolution. *Systems Research and Behavioral Science* vol. 6, no. 3, 1989, pp. 187-198.

Williams, R.J.P. and J.J.R. Frausto da Silva. 1996. *The Natural Selection of the Chemical Elements*. Oxford University Press.

CHAPTER OPENING QUOTES FROM HERACLITUS

(Translations by Burnet, 1930)

Chapter 1   **Energy In All Things**
It is wise to hearken, not to me,
but to my Logos, and to confess
that All Things are One.

Chapter 2   **A New Science of Order**
Wisdom is one thing. It is to know the
thought by which all things are steered
through all things.

Chapter 3   **Levolution in Scientific Context**
Though this Logos is true evermore, yet men
are as unable to understand it when they hear
it for the first time as before they have heard it
at all.

APPENDIX B
THE LAWS OF ENERGY

1. **The Conservation Law** – Energy cannot be created or destroyed, and is conserved in every reaction and process.

2. **The Entropy Law** – Energy spontaneously flows downward from higher to lower levels of potential, and moves to eliminate gradients of potential.

3. **The Absolute Zero Law** – Energy cannot be completely eliminated from any system or location. Energy is everywhere.

4. **The Maximum Entropy Production Law** – Energy in its entropic flow will allocate itself to the available pathways that allow it to flow downward in potential at the fastest rate possible, within constraints.

APPENDIX C
THE LAWS OF FUNCTIONAL ORDER

1. A **Dissipative Structure** is a morphodynamic attractor state representing a form or pattern of energy flows that maximize energy flow through the structure, and maximize its entropy production. They perform three universal functions: (1) energy capture from the environment, (2) energy use, including use in producing and maintaining their own functional order, and (3) entropic energy degradation and dissipation to the environment. Formation of this important type structure is the only mechanism known for thermodynamically compliant creation of Functional Order in nature.

2. **Thermodynamic Natural Selection** – When energy flows among a population or set of discrete structures, these discrete entities represent available energy pathways. Variation in their characteristics govern the pathway selection and energy allocation process inherent in the Maximum Entropy Production Law. Energy will differentially allocate itself to those discrete entities that transfer the most energy, resulting in differential thermodynamic stability and survival. This allocation process emerges at the population level as Thermodynamic Natural Selection among the variant structures.

This criterion underlies all other fitness criteria, because competition of any type results in differential rates of energy capture, and so differential rates of survival. Thermodynamic Natural Selection does not depend on reproduction or generations.

3. **Universal Thermodynamic Evolution** – Thermodynamic Natural Selection is a mechanism of adaptive change that applies universally to populations of variable holosystems, and some dissipative structures. Evolutionary change results in adaptive modifications of traits and generally increased fitness, to better match the challenges of a particular environment, which may itself be changing, and the goal is always to maximize entropy production. The environment, however, is recognized to actually be a whole, of which the subject population of holosystems represent the parts. Universal Evolution essentially maintains the adaptation of parts to their wholes.

4. **Levolution** – A population of discrete dissipative structures or holosystems existing as independent wholes may evolve into the parts of a new, larger, and higher order whole, creating a new level of organization. The process of differentiation and aggregation into a new holosystem proceeds by the process of Levolution. It links the processes of evolutionary differentiation with the thermodynamic ideal of the holosystem structure model, which leads to functional integration of the differentiated parts. Like Universal Evolution, Levolution follows from the mechanism of Thermodynamic Natural Selection acting among a population of variant entities.

5. The **Holosystem Structure** is a further qualified dissipative structure. Holosystems are always composed of parts that are also dissipative structures (or smaller holosystems). New holosystems are, by definition, always whole-part dualities formed by the process of Levolution, which creates successive new levels of organization. Holosystems are characterized by their performance of the

dissipative structure functions; energy capture, use, and dissipation, plus three more functions related to the management of their parts. They (1) distribute energy to parts, (2) integrate the useful energy output (work) of parts, and (3) manage the entropy of parts.

6. **The Holarchic Structure of Nature** – The sequential creation of holosystems by the Levolution process energetically links the whole-part dualities of the universe into axes of serially nested holosystems, or a holarchic structure. The Axes of Functional Order are based on the energy forms, and branch at the point where gravity's influence separates from that of electromagnetism. The **Holarchy of Nature** is the whole set of structures that have been built to help degrade the **Monodyne of Energy** in accord with the entropic goal and purpose of energy itself.